The Farm as Natural Habitat

The Farm as Natural Habitat

Reconnecting Food Systems with Ecosystems

Edited by

Dana L. Jackson and Laura L. Jackson

ISLAND PRESS

Washington • Covelo • London

ISBN 1-55963-846-X (cloth) — ISBN 1-55963-847-8 (paper)

Printed on recycled, acid-free paper 

Library of Congress and British Cataloging-in-Publication Data available.

Manufactured in the United States of America
09 08 07 06 05 04 03 8 7 6 5 4 3 2

Contents

Acknowledgments

Both of us wish to thank the group of people who came together at the Leopold Shack on June 12–14, 1998, for a seminar on agroecological restoration, which led to the conception of this book. They included Dave Andow, George Boody, Nina and Charles Bradley, Brian DeVore, Tom Frantzen, Paul Gruchow, Tex Hawkins, Buddy Huffaker, Rhonda Janke, Paul Johnson, Nick Jordan, Cheryl Miller, Stephanie Mills, Dave Minar, and Judy Soule. Thanks also to the Aldo Leopold Foundation for hosting that meeting at the Leopold Shack.

In addition, Mark Gernes, Doug Romig, and Holly Winger contributed ideas to an earlier meeting on biodiversity and sustainable farming on June 17, 1995.

Dana's Acknowledgments

I am grateful for the support and patience of my colleagues on the staff of the Land Stewardship Project (LSP) and for the encouragement of its board of directors. I especially appreciate LSP's executive director, George Boody, communications coordinator Brian DeVore, and board member Cheryl Miller for contributing chapters to the book. I thank George for being behind this project from the beginning. Brian DeVore was an invaluable advisor, editor, and "ear" from conception to finish. Cheryl restarted me when I needed it.

I was inspired to work on this book when I became acquainted with the six farm families who were members of Land Stewardship Project's Biological, Social and Financial Monitoring Team: Daniel and Muriel French, Ralph and Geri Lentz, Dave and Florence Minar, Mike and Jennifer Rupprecht, Joe and Marlene Finley, and Art and Jean Thicke. In visits to four of these farms for field days and pasture walks, I learned about holistic management, management intensive rotational grazing, and the connection between good farming and biodiversity. Many thanks to these farmers for helping us ground this book in real life goals and experiences.

Thanks to Paula MacKay for contributing the subtitle of this book.

Nina Bradley's encouragement meant a lot to me during the production of the book, as did that of Kay Adair and Sister Mary Tacheny. I thank my sister Cleo Richards and my daughter Sara Miller for their confidence that Laura and I could and would finish the book.

This work was partially supported by funds from the Pew Scholars Award that I received in 1990 and the Compton Foundation and the Foundation for Deep Ecology.

Laura's Acknowledgments

I wish to thank farmers Gene Flatjord, Tom and Irene Frantzen, Laura Krouse, Jan Libby, Michael Natvig, Dan Specht, Rich and Al Steffen, and Matt and Diana Stewart. Each has graciously helped me to understand their farms and how a farmer views his or her land. Any misrepresentations of their ideas, however, are my own responsibility.

I have benefited immensely from discussions with students Terrance Loecke, Brook Herman, Michele Smith, Angie Miller, Dave Williams, Dawn Keller, Judie Krebsbach, and many others who have taken my Applied Ecology and Conservation course over the years. Their stories have contributed greatly to my understanding of Iowa.

This work was partially supported by a Professional Development Leave grant from the University of Northern Iowa. I am grateful for all forms of support that the Department of Biology has given me. Some of the research described in my chapters was supported by the Fund for Rural America and the Leopold Center for Sustainable Agriculture.

Thanks to valued colleagues and friends Mark Grey, Daryl Smith, Paul Whitson, Ed Brown, Kirk Henderson, Ken Nuss, Laura Walter, Michael Walter, Jill Trainer, Dennis Keeney, Elizabeth Gilbert, Walter Goldstein, and Laura Paine for their useful insights and patient listening. Don Falk and David Andow provided helpful comments on drafts of my chapters.

This book has benefited greatly from discussions with members of The Pine Lake Group, consisting of Tom Richard, Clare Hinrich, Matt Liebman, Laura Merrick, Mike Bell, Diane Mayerfield, Diane Debinski, and James Pritchard.

Hilda Ostby, a treasured part of our extended family and community, cared for daughters Nettie and Ada while my husband and I were at work.

My husband, Kamyar Enshayan, has supported and encouraged this project in every possible way, and I am forever grateful.

I would last like to thank my father, Wes Jackson, for his example of passion and intellectual drive, and for his insistence on an ecological-evolutionary worldview for agriculture. This book is in part a response to his vision, and I hope, a worthwhile contribution to it.

Foreword

At a gathering of farmers in 1980, I was impressed with a comment from an Iowa farmer who had never used pesticides or herbicides on his land and whose farm was healthy, functional, and financially stable. His comment was that the government seldom acknowledges good farmers, either financially, politically, or socially.

It is important that we give appropriate support to our independent farmers. We need to raise awareness of the common values shared among farmers and environmentalists—values of satisfaction in owning land and being part of the community, values of beauty in native plants and animals, values in diversity and health of the land. As Paul Johnson has said clearly, "Conservation is not just about building another terrace, it's sharing the land with 100,000 other species."

The concept of *The Farm as Natural Habitat* presents the positive effects of good farming—benefits to the farm family, to the community, and most importantly to the land. These essays help shape our understanding of stewardship in agriculture. Although the remedy for a troubled landscape lies with those who manage the land, this collection provides suggestions for the responsibility of any citizen for land and conservation.

In 1948, Aldo Leopold wrote, "There are two spiritual dangers in not owning a farm. One is the danger of supposing that breakfast comes from the grocery, and the other that heat comes from the furnace." With today's fast-growing, urban society, we have almost lost track of these facts.

If we don't work hard and work together to improve farmers' profitability and secure a land base for farming in the face of urban development, many of us will live in sprawl, drive through sprawl, and work in sprawl. There won't be many farmers left, and the countryside will be dominated by the hilltop mansions of the few wealthy enough to escape.

Farmers have contributed a lot to conservation and biodiversity, as well as putting food on our plate each day. Aldo Leopold spoke clearly that we have to make a living from the land, that we all need shelter, clothes, and food. But he also realized that we need a great deal more if we are to lead sane and honorable lives; we need beauty, community, and purpose; we need "spiritual relationships to things of the land."

In 1940, Leopold wrote: "Conservation is a state of harmony between men and land." Such a state of harmony would be nurtured by the concept of the farm as natural habitat.

NINA LEOPOLD BRADLEY
Co-founder and Board member
Aldo Leopold Foundation

The Farm as Natural Habitat

Introduction

Laura L. Jackson and Dana L. Jackson

Both of us moved to the upper Midwest in 1993: Laura to Iowa and Dana to Minnesota. As we expected, large-scale, specialized agriculture dominated the landscape, and there were the usual problems—habitat loss, erosion, and water polluted with sediments, excess nutrients, and pesticides. But things were changing, and not for the better. Rows of long white buildings, each holding about a thousand hogs, began to appear on one farm after another. It was happening so fast no one could keep track of the numbers. Despite the unholy stench, the rancorous opposition from neighbors, and the manure spills, the buildings kept going up. Citizen meetings were held. Counties tried to pass ordinances that were subsequently struck down by the state courts. "This is the way we will raise all livestock now and in the future—it's inevitable" was the message from animal scientists, commodity organizations, and agribusiness leaders. It was as if animals were machines and the land grew the fuel to run the machines—technology could solve all problems that might arise.

At the same time, we were both beginning to meet family farmers with diversified cattle, hog, and grain farms in our respective states. All of them raised livestock on pasture and most rotated several crops with other land in hay and grass. The course of Dana's work took her to farm field days sponsored by the Land Stewardship Project and the Sustainable Farming Association of Minnesota. Laura's research and interests took her to field days held by farmers belonging to a group called the Practical Farmers of Iowa. She began to work with a few of them on the establishment of warm-season grasses and other prairie plants in grazing systems. We began to see that there were other farming patterns on the land and that there were farmers increasing biological diversity and improving water quality while staying economically viable. Why should people accept factory farms as the model of the future when there were farms like these that offered benefits to society rather than costs?

The Farm as Natural Habitat is about the connection between the grocery list and the endangered species list, between farming and nature. Many people are only vaguely aware that their food is not produced on Old MacDonald's Farm but in a sterile landscape of row crops drained by ramrod-straight, silt-laden streams and interspersed with meat factories housing thousands or millions of animals. A few consumers are concerned about the rising numbers of genetically altered crops, about the treatment of animals, and about the effects of farm chemicals in drinking water. But for most, farm country is not where you find nature anyway. Why bother, when it is so topographically challenged, so boring? If you want to enjoy nature, go to a park.

Although the average citizen may not often think about farming and nature together, professionals have taken a few steps in this direction. Knight (1999) and Norton (2000) each contributed editorials to *Conservation Biology* pointing out that the profession's focus on public lands has been too narrow. After all, farming is a major cause of habitat destruction for 38 percent of the 1,207 species on the U.S. federal endangered species list harmed by habitat degradation (including endangered, threatened, and those proposed for such designation). Agriculture leads the list of causes—neck and neck with commercial development (35 percent) and well above grazing (22 percent), logging (12 percent), and dams (17 percent; Wilcove et al. 1998). Despite the recent heads-up, however, few *Conservation Biology* articles concern themselves with agriculture, and those that do are usually set where agriculture is expanding into forests.

The Ecological Society of America (ESA) has likewise considered the importance of land use for nature conservation. The ESA published an extensive paper in *Ecological Applications* (Dale et al. 2000) that bravely outlined ecological principles for managing the land, including private land. However, because it assumed some monolithic "land manager," presumably of governmental origin (Cronon 2000), its results had little realistic application to agricultural policy.

Conservative social critic Dennis Avery has been thinking about farming and nature. He contends that the intensification and industrialization of agriculture in the northern, developed world—and sacrifice of nature there—will cause more food to be grown per acre, thus protecting other land elsewhere from agricultural development (Avery 1995). Pesticides, fertilizers, and varieties produced with the help of biotechnology are thus the saviors of nature in his view. This perspective has been enthusiastically promoted by a vice-president of Monsanto, Robert Fraley (Horsch and Fraley 1998). In contrast, a few U.S. agronomists and ecologists are beginning to calculate the costs of declining agricultural diversity (CAST 1999).

The World Conservation Union (IUCN) has been thinking about farming and nature in developing countries but has come up with a decidedly different conclusion than Avery's. A new report entitled *Common Ground, Common Future: How Ecoagriculture Can Help Feed the World and Save Wild Biodiversity* teams up IUCN with Future Harvest, a nonprofit organization that builds awareness and support for food and environmental research. Jeff McNeely and Susan Scherr (2001) propose a mixed strategy: increase agricultural production on currently farmed land, enhance wildlife habitat on farms, establish protected areas near farming areas, and mimic natural habitats within farming systems. They call for more ecological research in the service of agriculture and for government policy that rewards farmers' conservation efforts. The World Bank has also begun to address the problem of biodiversity and agricultural intensification (Srivastava et al. 1996). Without ignoring North America and Europe, both publications rightly focus on developing nations where human populations are growing fastest.

Until recently, large environmental organizations have mostly been concerned about protecting nature from agriculture, and they have lobbied Congress and state governments for stricter regulations on farming to protect water quality and endangered species. The widening discussion on "multi-functional" agriculture in Europe and "multi-benefit" agriculture in the United States has engaged their interest in collaborating with sustainable farm groups to promote nature-friendly farming in the 2002 farm bill.

Meanwhile, hamburgers and fries have been on the best-seller list. Eric Schlosser's *Fast Food Nation: The Dark Side of the All-American Meal* (2001) examines the implications of America's quintessential meal for livestock farmers, meatpackers, rural communities, children's health, and the treatment of farmland. The enormous buying power of the fast-food industry (the average American consumes three burgers and four orders of fries a week) has helped to move cattle away from grazing on pastures, to fattening on corn in a huge feedlot. More land must be plowed and planted, fertilized, sprayed, and irrigated to feed cattle. His other observations on everything from teen labor to food engineering to the suburban landscape show the broad social impacts of this food system.

The place we call home, the Midwest, the Breadbasket, and (a little too charitably, we think) the Heartland, is an ecological sacrifice area. Disparaged for its flatness, the Grain Belt is a notorious bore. Although big Midwestern cities are few, they do have interesting ethnic food, museums, art, music, and theater. What visitors really object to, we suspect, is the region's utter devotion to growing crops. Much as we protest that "really, once you get off the highway, it's quite nice," the visitors' objectivity cannot be denied. Unlike the visitor, however, we know that it was not always thus.

Although our perspective is Midwestern, few agricultural areas of the United States, Canada, and other developed countries are exempt from the model of agriculture practiced here. Corn is grown in huge fields, heavily fertilized, and sprayed with pesticides not only in Iowa, but also in eastern Colorado, west Texas, Pennsylvania, Vermont, and Europe. Millions of hogs are fed that corn in climate-controlled buildings in the upper Midwest, Utah, Oklahoma, North Carolina, Poland, and the Netherlands. Other crops with their own history of environmental damage (wheat, sunflowers, sugar cane, potatoes, cotton, vegetables, fruits) dominate other regions. And, worse, this model—of efficiency, specialization, homogeneity—continues to be exported to developing nations. What happens here, happens throughout the world.

This book is a reaction, in part, to the bullying notion that the agricultural landscape we see now in Iowa, with only one-tenth of 1 percent of its original vegetation intact, and home to some of the most nutrient-polluted lakes and streams in the United States, is inevitable. Many of the contributors to this book know farmers who are already making a solid living and improving the soil, water, and biological diversity on their land. *The Farm as Natural Habitat* stems from the conviction that the agricultural landscape as a whole could be restored to something better. The destruction of every last shred of nature is not a necessary compromise for the survival of the family farm or because of the need to "feed the world." In this book, we maintain that the trend toward sterile, industrialized agricultural is an unacceptable, unaffordable sacrifice, that it is far from necessary, and that we can help farmers reverse it to benefit nature conservation, rural communities, farm families, urban residents, and consumers.

Beyond reacting to real or perceived prejudice, this book is also a straightforward attempt to respond to and carry forward the insights of Aldo Leopold. In the Midwest, Leopold is not just a quotable icon of environmentalism but also a scientist whose observations on farming and nature were dead-on then and still accurate today. He wrote about the prairie ecosystem underlying current land uses, the incremental habitat loss caused by the industrialization of agriculture, the follies of land grant schools of agriculture (promoting exotic species and wetland drainage) and the awkward relationship between private land owners and government conservation programs (see chapter 5). Leopold's "The Farmer as a Conservationist," written in 1939, articulates the need for more skillful conservation based not on fear but on a "lively and vital curiosity" about the workings of ecology. He points out the hope and the potential of conservation in farm country, where most of the losses of species and ecosystem services have been avoidable mistakes, not born of necessity (Leopold 1991).

Although prescient in most respects, Leopold could not have predicted the changes in the global food system, from field to belly. He believed that "the landscape of any farm is the farmer's portrait of himself." Today the landscape of the farm is more like a portrait of Archer Daniels Midland, the global grain processor whose motto is "supermarket to the world." A small group of agricultural suppliers and processors have a huge influence on what farmers can grow and market—and their political power is enormous. Although change in the hearts of landowners is important, even more important is change in the national policy regarding food, agriculture and the environment. The aim of this book is to encourage policy change by providing examples of agriculture that could produce not only healthful food, but also ecosystem services and viable populations of almost all of the native species that once made this area home. No, we will not be able to sustain free-ranging herds of large herbivores and packs of wolves given current demands for food, but we can do better than the sacrifice area we see today.

Another goal of the book is to engage diverse groups of people, including those who work for conservation of biological diversity, and those who farm or work with farmers. The Wildlands Project's published goal (Noss 1992) to convert half of the continental United States, including half of the food-growing prairie states, back to wilderness reflected little concern for the livelihoods of farmers living there. People promoting sustainable agriculture have not made common cause with endangered wildflowers or freshwater mussels. This too can change. In 2000, wildlands proponents and sustainable agriculture activists founded the Wild Farm Alliance to forge a new coalition based on common interests. This organization hopes to build a network of farmers, conservationists, and consumers all promoting a kind of farming that helps protect and restore wild nature.

Sustainable Agriculture, Agroecological Restoration, and Ecosystem Management

Many farmers, environmentalists, and social activists called for a reform of industrial agriculture in the early 1980s. This started the sustainable agriculture movement, which began with a strong critique of the farming practices used by industrial agriculture that caused soil erosion and water contamination. The movement complained that large farms, big equipment, and livestock confinement grew larger at the expense of family farms, good stewardship, and animal husbandry. The land grant colleges of agriculture were indifferent to the plight of family farmers going bankrupt but fascinated with improvements in technology to create higher yield and serve large farming operations. Initially, they were uninterested in low-cost practices based on alternatives to purchased inputs. So farmers and nonprofit organizations began experimenting with alternative systems of farming on their own, and in almost every state an organization of sustainable farmers began to learn, teach, and promote the alternatives. Although industrial agriculture hasn't been reformed, many farmers are doing things differently and making a living on the land. Some land grants have even established small programs in sustainable agriculture. But something else is happening. The sustainable farmers are restoring a relationship between farming and the natural world that welcomes greater biodiversity and the use of free ecological services in their operations. Out of sustainable agriculture is emerging new possibilities for what we like to call *agroecological restoration*.

How can we possibly include "agro" and "ecological restoration" in the same expression? Typically, what comes to mind when one thinks of ecological restoration is the image of people seeding wild plants. But farms must grow domesticated cereals, legumes, and livestock where forests and prairies and wetlands once stood. So how can we restore nature to farms? The answer comes in part from a expanded understanding of "nature." Nature includes soil carbon/nitrogen ratios, the distribution of soil aggregate sizes and the pore spaces they create, and the movement of molecules through soil profiles, through the food web of soil biota, and into groundwater. It is the way water moves across and through a landscape after precipitation and the way sediment moves in a stream. Nature is the vertical structure of vegetation, from ground-hugging lichens to successively taller plants, and their calendar of photosynthetic activity. Nature is the way migratory birds cross continents, resting and feeding in places that are individually short-lived but collectively ageless, like sandbars or river backwaters.

Ecological restoration focuses on basic ecosystem processes and disturbance regimes as much as it does on indigenous species. Although agriculture will never look entirely like "nature" in terms of species composition, we could ask it to be "natural" in terms of other ecosystem processes. For instance, we know that infrequently it can rain 14 inches in twenty-four hours in late June, when fields of corn and soybeans are not completely covered with vegetation. The resulting erosion is catastrophic. Our methods of growing food should be resilient, as the original vegetation was, to such extreme but entirely normal events.

This book also has ties to the idea of "ecosystem management," the new cornerstone and philosophy of federal natural resource agencies (Christensen et al. 1996). As we will suggest in Part III, ideas of ecosystem management are particularly relevant to agriculture, although they have yet to be applied.

The Flow of the Book

The book is divided into four parts. Part I lays out the problem as we see it: modern industrialized agriculture has become an ecological sacrifice zone impacting not just agricultural regions but also the Gulf of Mexico. Furthermore, unless we do something, the tiny preserves of biological diversity will continue to degrade within these landscapes due to their size, isolation, and vulnerability to larger-scale processes. However, positive experiences with farmers tell us that agriculture does not have to be a sacrifice zone, and rural areas do not have to be unhealthy places to live.

Part II introduces readers to some of the farmers who have inspired us to reject our prejudices. These are people who have belied the notion that success—indeed, economic survival—comes at the expense of a biologically rich landscape. We tie these farmers and their accomplishments to the ideas of Aldo Leopold, who more than fifty years ago articulated the problem of merging nature and agriculture. Farm journalist Brian DeVore explores not just the what and how but also the why—why is it that some farmers have been moved to expand their definition of the farm to include biological diversity? What does it mean to them and how far will they go?

Part III explores some concepts that may someday amount to ecosystem management of agricultural landscapes. Right now we are very far away from knowing how that could work on privately owned land, but the examples in these chapters—from northern California to southern Texas, from Ohio to Great Britain—are a good start. Chapters in Part III also ask some theoretical questions. What does it mean to restore nature in the context of an intensively managed landscape? How are biodiversity and production related—positively or negatively? What are the limits of mutual benefit and cooperation when conservationists and farmers occupy the same space? How do natural resources conflicts such as water scarcity and disease epidemics constrain our ability to restore nature on farms?

Part IV outlines steps that we need to take to begin meaningful ecological restoration of our agricultural landscapes. This restoration will occur at many different levels. It must occur at the level of the farm, where individuals balance the improvements and changes they'd like to make with the realities of cash flow. It must occur at the level of rural communities and their relationships with environmentalists and regulators in urban areas. It must occur within individuals, as they look within themselves to find concern for all land, not just wilderness. And it must occur at the kitchen table and in the grocery store, the food coop, and the farmers market. Finally, we suggest that changes to the federal farm program must be made in order to make the "public good" a desired outcome in the management of agricultural land.

This book does not attempt to be an exhaustive treatment of the subject. For instance, we have not attempted to review all the pitfalls of modern agriculture—that has been done well elsewhere. Nor can we hope to present the full array of techniques and strategies that are already available to restore nature to farms. Their absence in this book does not connote disapproval.

We have not included a discussion of the Missouri and Mississippi Rivers, both damaged by sedimentation, dams, and channelization. (The Missouri River was recently named in 2001 by American Rivers as the nation's most endangered—the pallid sturgeon, the piping plover, and the interior least tern are expected to become extinct if dam operations are not modified.) The rivers are managed primarily to accommodate the barge industry. Barges carry mainly agricultural products—fertilizers and pesticides upriver, agricultural commodities downriver. Certainly, principles of ecosystem management should be applied to these rivers, and that case has been made elegantly elsewhere (Sparks et al. 1998, Galat et al. 1998). Of course, improvements in the nutrient cycling, hydrology, and land cover of the immense territory drained by these great rivers will contribute a good deal to their restoration as well. A comprehensive vision of the entire region would not separate rivers from terrestrial issues, and indeed the Mississippi, its tributaries, and the zone of hypoxia in the Gulf of Mexico are discussed by several contributors to this book.

Finally, a word about the different voices in this book. What they have in common is experience working in association with farmers; most of the authors regularly get out on farms and talk to farmers. Their professions, however, are quite diverse. Contributors include an agricultural journalist with close ties to farmers, academic ecologists and agronomists, agency personnel trained in ecological inventory and natural resource management, and staff members of nonprofit organizations devoted to sustainable agriculture. Their styles of expression are equally diverse and, we believe, up to the task.

References

Avery, D.1995. *Saving the Planet with Pesticides and Plastic: The Environmental Triumph of High-Yield Farming*. Hudson Institute, Indianapolis.

Christensen, N. L., A. M. Bartuska, J. H. Brown et al. 1996. "The Report of the Ecological Society of America Committee on the Scientific Basis for Ecosystem Management." *Ecological Applications* 6:665–691.

Council for Agriculture Science and Technology (CAST). 1999. *Benefits of Biodiversity*. Task Force Report no. 33. Ames, Iowa.

Cronon, W. 2000. "Resisting Monoliths and Tabulae Rasae." *Ecological Applications* 10:673–675.

Dale, V. H., S. Brown, R. A. Haeuber, N. T. Hobbs, N. Huntly, R. J. Naiman, W. E. Riebsame, M. G. Turner, and T. J. Valone. 2000. "Ecological Principles and Guidelines for Managing the Use of Land." *Ecological Applications* 10:639–670.

Galat, D. L., L. H. Fredrickson, D. D. Humburg, K. J. Bataille et al. 1998. "Flooding to Restore Connectivity of Regulated, Large-River Wetlands." *BioScience* 48:721–734.

Horsch, R. B., and R. T. Fraley. 1998. "Biotechnology Can Help Reduce the Loss of Biodiversity." Pp. 49–65 in *Protection of Global Biodiversity: Converging Strategies*, edited by L. D. Guruswamy and J. A. McNeely. Duke University Press, Durham, N.C.

Knight, R. L. 1999. "Private Lands: The Neglected Geography." *Conservation Biology* 13:223–224.

Leopold, A. 1991. "The Farmer as a Conservationist." Pp. 255–265 in *River of the Mother of God and Other Essays*, by Aldo Leopold, edited by J. B. Callicott and S. L. Flader. University of Wisconsin Press, Madison.

McNeely, J. A., and S. J. Scherr. 2001. *Common Ground, Common Future: How Ecoagriculture Can Help Feed the World and Save Wild Biodiversity*. Report published electronically by the World Conservation Union and Future Harvest. Available from http://www.futureharvest.org, accessed May 8, 2001.

Norton, D. A. 2000. "Conservation Biology and Private Land: Shifting the Focus." *Conservation Biology* 14:1221–1223.

Noss, R. F. 1992. "The Wildlands Project: Land Conservation Strategy." Pp. 233–266 in *Environmental Policy and Biodiversity*, edited by R. E. Grumbine. Island Press, Washington, D.C.

Schlosser, E. 2001. *Fast Food Nation: The Dark Side of the All-American Meal*. Houghton Mifflin, Boston.

Sparks, R. E., J. C. Nelson, and Y. Yin. 1998. "Naturalization of the Flood Regime in Regulated Rivers." *BioScience* 48:706–720.

Srivastava, N., J. H. Smith, and D. A. Forno, editors. 1996. *Biodiversity and Agricultural Intensification: Partners for Development and Conservation*. World Bank, Washington, D.C.

Wilcove, D. S., D. Rothstein, J. Dubow, A. Phillips, and E. Losos. 1998. "Quantifying Threats to Imperiled Species in the United States." *Bioscience* 48:607–616.

Part I

Agriculture as Ecological Sacrifice

Hamburger and ham sections in the meat cases of Krogers and Safeways, Super Walmarts and Super Targets, are kept full—all across the United States. Drivers of delivery trucks unload hamburger patties and ham slices to fill the freezers of Hardees, McDonalds, and Wendys along every major highway. Environmentalists worry about the excessive use of paper and Styrofoam, forests turned into dishes and fossil fuel into white blobs, found on every human path from Main Street to nature trail. Nutritionists worry about lack of variety in the diets of overweight teenagers who buy triple-patty hamburgers with cheese *and* large fries for dinner one evening, and then get the same meal in the school lunchroom the next noon, and again in front of the TV at home that evening.

The land that brings this fatty diet to Americans suffers from glut too, a surfeit of chemical nitrogen and manure applied to crops and leached into groundwater, soils gorged with phosphorus, and streams belching hog feces. Once covered with many kinds of forbs, grasses, and trees, and later a diversity of crops, the land now sours on continuous corn and soybeans.

But think of the convenience of stopping outside of Des Moines for an egg-ham-and-cheese McMuffin breakfast before we head north through the boring agricultural landscape to spend a week with nature in Voyageurs National Park.

Expectations for experiences with nature in farming country are very low these days in the context of modern agriculture. This land has been dedicated to producing high yields of feed grains and millions of livestock. Hay meadows with wildflowers, tree-lined fishing creeks, marshes noisy with waterfowl, clear springs making watering holes for cattle on pasture—these have been sacrificed, along with the less-visible ecosystem processes that undergird a healthy land. In the dying small towns, clerks apologize to tourists attracted to their Main Street's country craft shops in lovely old Victorian houses as they catch the strong stench of hog manure between car and shop.

Agriculture as ecological sacrifice is generally accepted (though not acknowledged out loud) in U.S. Department of Agriculture offices where farmers go to fill out farm program papers, or in offices of the county extension. Why is this happening? How have we come to believe that it's normal for country folk to go to a city park for scenic beauty, fresh air, and swimming? The three chapters in Part I provide background for this state of affairs but do not accept it as permanent. The whole point of this book is to reject the notion of agriculture as ecological sacrifice and restore nature in agriculture.

In chapter 1, "The Farm as Natural Habitat," Dana Jackson acknowledges that the rural countryside is becoming an industrial zone, but rather than accept that as its fate, she proposes a different vision based on principles of Aldo Leopold. We must work to make the farm a natural habitat, not a factory. She's not proposing the planting of natural habitats on farms but rather making the farm itself into a natural habitat for humans who live there, for domestic crops and livestock, and for the wild or native plants and animals that can be accommodated. Some farmers have been led to this joyous turnaround through adoption of management intensive rotational grazing (rather than continuous grazing of pastures or a total grain diet for livestock) and a planning process for their farm that has led them to manage the land's natural resources with greater attention, while earning a good living too.

In chapter 2, "Nature's Backlash," Brian DeVore gives a picture of what is going wrong with industrial agriculture. In spite of its history of successfully ignoring natural processes and overcoming nutrient deficiencies and weed and insect infestations with specialized technology and equipment, today's solutions to problems aren't always working. DeVore shows that hog manure spills and the indomitable wheat rust in the Red River Valley are problems that aren't being solved because nature and people are more complex than the industrial model assumes.

Laura Jackson writes in chapter 3, "The Farm, the Nature Preserve, and the Conservation Biologist," about the problem of protecting natural areas within an industrial agricultural landscape, which she illustrates by describing a 4-acre prairie surrounded by fields that she and her husband lease in order to protect it. But the big question is: why do conservation biologists consider it acceptable for land in farms to be an ecological sacrifice zone? She suggests some reasons why biologists have avoided farming landscapes and stuck to public lands, but she nevertheless maintains that conservationists need to devise ways to reconcile the needs of wild nature with farms.

Chapter 1

The Farm as Natural Habitat

Dana L. Jackson

"We should be having our summer board meeting on a farm. It's really beautiful at my place now." When Dan French brought this up at the annual meeting of the Land Stewardship Project's board of directors, everyone nodded in agreement. Why didn't we think about scheduling the meeting there? I had spent a few days at the French farm several summers ago, sitting under a tent listening to instructors in the holistic management course held there and looking at Muriel's flower garden and the black-and-white dairy cows in the pasture beyond. A light breeze brought us the fragrance of green alfalfa from the barn where Dan's son was unloading bales. The class walked down to the creek, where we talked about the water cycle and how to judge water quality by observing the kinds of insects and fish in the water. The gravel on the creek bottom sparkled in the clear water. There was a flurry of birds and birdsong in the taller grass of a pasture section that hadn't been grazed for awhile.

When you drive up to the French farm, it looks like an interesting place, with its barns and outbuildings, vegetable and flower gardens, and shady picnic area. It doesn't look like those farmsteads one often sees in the Corn Belt, where the house and a machinery shed or two seem to just stick up out of a corn field, as if the owners had planted every inch they could on the place. Actually, if you drive on Interstate 35 between Saint Paul and Des Moines, you don't even see many houses. The landscape in July seems to be covered just with corn, a seemingly endless monotony of green stalks broken occasionally by shorter bushy soybeans.

It's hard to imagine what it must have looked like when Europeans first settled the Midwest, when it was a wilderness covered with prairie, forests, clear streams, and herds of buffalo. Too quickly it became dominated by agricultural uses interrupted by a few patches of prairie or woods around lakes or rivers that harbored remnants of natural habitat. Some prairie plants survived in pastures and meadows until they were replaced by fields of corn and soybeans in the last part of the twentieth century. Then animals were moved into barns and feedlots, fences came down, and habitat edges disappeared. It only took about 150 years to reduce biological diversity on this landscape to a numbing sameness.

It is no surprise that people passionate about wildlife and the preservation of natural habitats have concentrated on protecting other places, those dramatic expanses of land where more of the original landscape remains, such as the Boundary Waters Canoe Area in northern Minnesota, the rugged mountains of Colorado and Montana, and roadless areas in Alaska. Such conservationists have accepted the agricultural Midwest, especially the Corn Belt, as a sacrifice area, like an open pit iron mine, or an oil field, where we mine the rich soil and create toxic wastes to extract basic raw materials. But the environmental impacts of this kind of mining are not confined to farming country. No nature preserves within its watersheds or wildlife area downstream on the Mississippi River can be adequately protected from farming practices that simplify ecosystems to a few manageable species and replace ecosystem services with industrial processes.

People who live in rural areas or urban residents who drive through them may not know that they are seeing a biologically impoverished landscape, because they have no knowledge of its diversity before modern agriculture. Others may know or imagine what the land looked like with different kinds of crops, meadows, and livestock in pastures, but they accept its simplification because they are convinced that the main trends in agriculture can't be overcome. Agribusiness has successfully persuaded farmers, politicians, civic leaders, and even conservationists to believe that agricultural modernization leads to specialization and industrialization, and that financially viable alternatives are unavailable even though such modernization reduces the rural quality of life and harms the environment.

In this chapter, I will introduce an alternative vision for agriculture that defies the trends considered inevitable. It is a vision inspired by Aldo Leopold's writing that farming and natural areas should be interspersed, not separated, and by the farmer-members of the Land Stewardship Project, whose ways of managing farms have created a natural habitat for them, for their crops and livestock, and for the native plants and animals of the area. I will also describe two sustainable farming practices that currently are improving biological diversity on rural landscapes and showing the real possibility of this vision. Subsequent chapters describe these farmers and many others in more detail and discuss issues relevant to the vision and its feasibility. But first, let's look at the practices of mainstream industrial farming that render the countryside an ecological sacrifice zone.

Rural Lands as Industrial Zones

The loss of biological diversity was not the only environmental consequence of creating the Corn Belt. Soil erosion, depletion of water resources, contamination of groundwater and surface water from fertilizers and pesticides (Soule and Piper 1991), and a steady silt load in rivers are some of the consequences of so much tilled land. The sediment load in the Minnesota River at Mankato is equal to a ten-ton dump truck load moving by approximately every five and a half minutes (Minnesota Pollution Control Agency 1994).

The most serious environmental consequences are yet to come because of the growing consolidation in the livestock industry fueled by the abundance of cheap corn and soybeans. Each year an increasing number of poultry and hogs raised in the Corn Belt are not dispersed across the countryside on independent farms but are instead concentrated in large operations. Hundreds of thousands of chickens and tens of thousands of hogs are confined in buildings, creating huge quantities of manure that pose serious environmental risks to ground and surface water, as Brian DeVore describes in Chapter 2. Hydrogen sulfide fumes in the stench emitted from the operations have sickened neighbors. People don't want to live close to these hog factories or visit relatives close to hog factories. The once rich prairies that became bucolic communities are now industrial zones, suitable for "neither man nor beast."

Dairy farmers also feed the bounteous harvest of the Corn Belt to cattle confined in barns and milked three times a day. Dairy operations with one thousand to two thousand cows are replacing traditional family-sized farms with 100 or fewer cows. They manage large quantities of manure the same way as hog factories do and present the same risks to water quality. Travelers through Wisconsin's wooded hill lands graced with small dairy farms in the valleys may be unaware of how this landscape will change if consolidation continues in the dairy industry and four dairy farmers go out of business each day in the state as they did between 1992 and 1997 (USDA 1997). Where large-scale dairies replace small ones, the scenes of black-and-white cows grazing on green pastures and moving in line to and from red barns are being replaced by fields of corn and soybeans with nary a cow in sight.

Factory livestock operations have popped up like mushrooms across the entire Midwest and Great Plains. They have also grown rapidly in southern states and are emerging everywhere state laws are weak and local communities naively believe the industry's forecasts for economic development. California led the way with its one thousand cow dairies and became the leading milk producer in the country; as a result, departments of agriculture in traditional dairy states are promoting California-style dairying. Agricultural economists encourage farmers to expand their operations to be efficient and convince them that all dairy cows and pigs, like poultry, are going to be raised in large-scale confinement operations in the future. It's inevitable.

This mantra of "it's inevitable" is happily chanted by the corporate processors of pork that benefit from large supplies of cheap hogs, and, sadly, this mantra is repeated by many farmers. Some of them borrowed heavily to expand and build hog confinement buildings, and when pork prices plunged to an historic low in the 1990s, they went bankrupt. The huge packing plants that encouraged industrial production prospered and consolidated into even larger corporations through mergers (Heffernan 1999).

We are seeing rural landscapes all across the United States changing for the worse because farmers believe that further industrialization in livestock agriculture is inevitable and that they must "get big or get out." Some farmers incur staggering debt to increase the size of their operations, some form family corporations to share the costs of expansion, others invest in new buildings and technology to become contract producers for corporations, and some just leave farming and sell out to neighbors who want to expand. A house in the country isn't so romantic any more, because it might very well be within odor range of one of these hog expansions. Hay meadows and pastures with wildflowers and grassland birds are few and far between, and many streams running through fields have been cleared of trees and wildlife. If a family can't earn a living on the land, and it isn't a beautiful or healthful place to live, they might as well move to town. The land serves utilitarian purposes only, sacrificing natural values that once made it a home, not only for humans, but also for all kinds of creatures.

The disappearance of diversity in farming country has occurred steadily, mostly without notice or comment. Politicians and policymakers, the U.S. Department of Agriculture, land grant universities, and many farmers and rural people accept the loss of biological diversity on the land as a necessary cost of efficient high production. There is some nostalgia in older people for a favorite fishing or swimming hole on the creek of the farm on which they grew up, but farming is a business and you can't be sentimental about it. Most travelers aren't aware that many of the monotonous fields they see along the highways harbored wildlife in prairie pastures and hayfields as recently as the 1960s. They only know that if they want to see woods and prairies and wildlife, they must head for a publicly owned park or wildlife area where agriculture is not practiced.

Aldo Leopold and a Different Vision for Agriculture

Aldo Leopold, the Midwest's most famous conservationist, disapproved of the separation of natural areas from farming. To him it didn't make sense to protect forests in a special area and accept the absence of trees on agricultural land, when the farm was then left without the conservation benefit of erosion control and wind breaks. "Doesn't conservation imply a certain interspersion of land uses, a certain pepper-and-salt pattern in the warp and woof of the land use fabric?" he asked (Leopold 1991). Leopold believed that conservation efforts on certain parts of the land would fail if other parts were ruthlessly exploited. He wrote in the essay "Round River":

> Conservation is a state of harmony between men and land. By land is meant all of the things on, over, or in the earth. Harmony with land is like harmony with a friend; you cannot cherish his right hand and chop off his left. That is to say, you cannot love game and hate predators; you cannot conserve the waters and waste the ranges; you cannot build the forest and mine the farm. The land is one organism. (Leopold 1966)

Although Leopold knew that agriculture was becoming more industrialized and wrote about the dangers of a farm becoming a factory, he could not have imagined the enormous livestock factories in production today. The transformation of so many meadows, prairies, and wetlands into corn, beans, and hogs in Iowa, the state of his birth, and conversion of family-sized dairy farms into milk factories and corn fields in his adopted state of Wisconsin would astonish and grieve him. However, if someone told him about the zone of hypoxia in the Gulf of Mexico, seven thousand square miles depleted of marine life because of excess nutrients flowing down the Mississippi River from the Corn Belt, I doubt if he would be surprised.

It is understandable that people accept these trends as the destiny of agriculture if they cannot clearly see alternatives. But there is an alternative—another trend—that could produce a landscape of farms which are natural habitats rather than ecological sacrifice areas.

A strong minority of modern farmers, like Dan and Muriel French, have not turned their farms into factories nor abandoned their chosen profession but are instead leading agriculture in an entirely different direction. Their creative initiatives to make farming more economically sound and environmentally friendly are producing benefits for them, for society at large, and for the land. The trends of these models are toward independent farms supporting families and communities while restoring biological diversity and health to the land.

Using an ecological approach to management decisions, these farmers are restoring a relationship between farming and the natural world that improves the sustainability of both. This relationship makes the farm a natural habitat. It is a natural habitat for humans in that it is a healthful and aesthetic place to live and earn a living. The farm is a natural habitat for the crops and livestock because they are able to use ecosystem services for fertility and pest control rather than fossil fuel and man-made chemicals. And the farm is a natural habitat for native plants and animals, a refuge that encourages biological diversity along streams, in pastures, and along uncultivated edges.

Farming Practices for Natural Habitat

Farmers themselves don't talk about turning their farms into natural habitats. It happens as a result of the way that they choose to farm. Many farmers became interested in changing their practices in the 1980s, particularly during a period of low prices, high production costs and minuscule profits. A number of newly formed farming organizations around the country helped them lower their use of purchased inputs, such as chemical fertilizer and pesticides, and develop more environmentally friendly practices. For example, the Land Stewardship Project (LSP) in Minnesota began to hold workshops and field days about the practice of *management intensive rotational grazing*. This involves dividing a pasture into sections or "paddocks" with electric fences and allowing the animals to graze each area intensively for a short period of time before moving them on to another area. In conventional grazing, livestock roam freely in an open pasture, often overgrazing some areas and causing erosion.

Management intensive rotational grazing roughly mimics grazing patterns of migrating buffalo herds that preceded European settlement on the plains and prairies, but domestic livestock return to graze an area much sooner than did buffalo. The length of time that animals graze a particular paddock usually depends upon the rate of recovery of the forage after grazing and its nutritional value, which requires farmers or ranchers to become attentive observers of their pastures and all that is growing there.

A group of farmers wanted to know how they could tell whether the switch to management intensive rotational grazing was making their farms more sustainable. In response, Land Stewardship Project established a biological, social, and financial monitoring project, conducting research on six diversified livestock and dairy farms that used management intensive rotational grazing. The project team that worked together for three years included university researchers and state agency staff in addition to the six farmers and LSP staff. To conduct biological monitoring, researchers helped the six farmers collect biological, physical, and chemical soil quality data from sixty plots and make observations about pasture vegetative species and ground cover. They sampled wells and kept precipitation records. The farmers learned to survey their land for breeding birds, frogs, and toads, and they helped fisheries scientists survey streams passing through four of the team farms and through one paired farm to analyze the effects of management intensive rotational grazing on stream banks and stream invertebrate and fish populations.

These farms were seen as natural habitats, not as ecological sacrifice areas. The farmers wanted to find out if the soil and the water quality in streams on their farms were improving, just as they wanted to know if their financial bottom lines were improving by cutting production costs. They were not accepting the "inevitable," that they must get big or get out.

The farmers in the monitoring project, and many others who have been constituents of the Land Stewardship Project, practice holistic management, a decision-making process based on goal setting, planning, and monitoring. This process was developed by Alan Savory, who founded the Center for Holistic Resource Management in Albuquerque, New Mexico, in 1984. Land Stewardship Project staff taught many holistic management courses throughout the Upper Midwest. They developed a research project to monitor the effectiveness of management decisions made by the six farmers who had taken the course and were making the switch from conventional grazing to management intensive rotational grazing.

Holistic management contains four elements that distinguish it from conventional farm management and provide managers with strong incentives to make environmentally sound decisions. First, as part of the goal-setting process, it directs managers to develop a long-term vision for how they want the landscape to look far into the future. Second, the model teaches basic recognition of ecosystem processes that farms are dependent upon: the water cycle, the mineral cycle, plant succession, and energy flow (Savory 1998). Farmers strive to understand these processes and harmonize their farming practices with them. For example, farmers can rely on nitrogen fixation in legumes and the recycling of nutrients in manure to provide fertility for fields. Third, holistic management places a high value on biological diversity both in crop systems and in areas on the land not used for farming. And last, practitioners consider the effect of any proposed action or choice of enterprise upon quality of life for the community as well as for themselves. They understand that their land is part of a larger whole and how they manage it will affect the landscape around them and the lives of people in the community. Holistic management has become an effective tool for those who want to be good stewards of the land and earn a living on it at the same time.

Though holistic management has been used on all kinds of farming operations, it was developed by Allen Savory in connection with rotational grazing. Farmers in the Upper Midwest often began using holistic management and management intensive rotational grazing approaches simultaneously. Cattle grazing on public lands in western states has been considered such a disaster by environmentalists that many have a negative view of grazing anywhere. However, at the landscape level in the Midwest and in parts of the Great Plains and the South management intensive rotational grazing provides visible environmental improvement in farming, especially where field crops have been converted to permanent pastures and livestock eat more grass than grain. Fewer acres of corn and soybeans also mean fewer applications of chemical pesticides, herbicides, and fertilizer, which decreases the potential for contamination of surface and groundwater. When corn and soybeans are replaced by perennial grasses, there is less soil erosion (Cambardella and Elliot 1992, Rayburn 1993).

Dairy farmers have widely adapted management intensive rotational grazing. Between 1993 and 1997, the number of Wisconsin dairy farmers using variations of this grazing method increased by 60 percent (ATFFI 1996). Milk cows on most conventional dairy farms are confined in "loafing barns" or corrals between milkings and are never allowed out to graze. On very large operations of 500–1,500 or more cows, feed is brought to the cows and all of their manure is pumped out of manure pits or scraped and hauled out of the barns to be spread on fields. Conventional dairy farmers work hard to produce the corn and alfalfa to feed the dairy herd, and capital costs for equipment and barns are high. In contrast, grass-based dairy farmers usually move cattle daily but claim that their work load and costs of production are much less because the cattle walk around in the paddocks, get most of their own food, and disperse their own manure (ATFFI 1996). With more feed produced in pastures, a farmer uses less machinery and fossil fuel (Rayburn 1993). Some grass-based farmers "don't have much iron," as they say, because they've sold most of the machinery they formerly needed for large fields of corn. With fewer acres planted for feed, they can share machinery with neighbors, employ custom harvesters to bring in their crops, or even buy feed from other farmers. For these dairy farmers, management intensive rotational grazing is a farming practice that benefits them as much as it benefits the land and the water.

Poultry and hog farmers also use management intensive rotational grazing. Hogs can be put on pasture to graze, at least for part of their food, and spread their own manure in the grass. Hogs can spend most of their time outdoors and farrow in pastures. Farmers in the Upper Midwest often combine outdoor and indoor production systems by bringing hogs into open-ended metal hoop buildings covered with canvas for the winter. Hogs bed in deep straw or corn stalks, which composts with their manure, warming the hogs in the process and producing nearly composted, dry fertilizer for the fields when the barns are cleaned. Manure is not a toxic waste in management intensive rotational grazing or hoop house production systems, and the cost to the farmer of handling it and the public for regulating it is little or nothing. In fact, overall production costs are so much lower that farmers can make a profit as long as they have fair access to markets (Dansingburg and Gunnink 1995) or sell cooperatively with other farmers or directly to consumers. If market prices are too low, farmers can use these hoop houses for other purposes, such as storing hay or machinery, which gives them a flexibility that producers trying to pay off the debt for a high-tech, single-use confinement facility don't have. Using management intensive rotational grazing and deep-bedded straw systems in hoop houses, farmers can take advantage of ecosystem services in providing animal feed and managing manure. These systems are efficient alternatives to the industrial production models for livestock and can compatibly exist alongside or as part of natural ecosystems.

The Benefits of Diversity

Diversified farms producing feed for their own livestock may rotate crops of alfalfa or other legumes, corn, soybeans, and small grains such as barley or oats, in contrast to conventional cash grain farms that rotate only corn and soybeans or grow corn with no rotation. For example, Jaime DeRosier employs a complex rotation of hay, wheat, barley, vetch, flax, buckwheat, corn, and soybeans on his large organic farm in northwestern Minnesota (DeRosier 1998). The Fred Kirschenmann farm in North Dakota rotates up to ten different grain or hay crops in three different rotations (Anonymous 2000). In all parts of the country, farmers are also planting several different kinds of grasses and legumes in their pasture mixes, planting fields in strips of several crops, intercropping one species with another (such as field peas with small grains), and using cover crops between plantings of major crops. In California, orchards, vineyards, and specialty crop farms have added cover crops and farmscape plantings to attract pollinators and other beneficial insects (CAFF 2000).

The benefits of biodiversity in agriculture were effectively laid out in a report with that title by a task force of the Council for Agriculture Science and Technology, co-chaired by ecologist G. David Tilman and geneticist Donald N. Duvick (CAST 1999). The report stresses the dependency of modern agriculture upon biological diversity and advocates greater attention to preserving diversity both in domesticated crops and livestock, and in the natural landscape.

The *Benefits of Biodiversity* also discusses the dependence of modern agriculture upon ecosystem services, such as pollination, generation of soils and renewal of their fertility, pest control, and decomposition of wastes. It acknowledges the importance of preserving biodiversity by protecting natural areas and proposes that we substantially increase the worldwide network of biodiversity reserves and preserve large blocks of land in native ecosystems.

This report was not produced by CAST for the purpose of rerouting agriculture from the direction trends are leading. However, if followed, just one recommendation would lead us toward a landscape of farms that are natural habitats:

> Increase the capacity of rural landscapes to sustain biodiversity and ecosystem services by maintaining hedgerows/windbreaks; leaving tracts of land in native habitat; planting a diversity of crops; decreasing the amount of tillage; encouraging pastoral activities and mixed-species forestry; using diverse, native grasslands; matching livestock to the production environment; and using integrated pest management techniques.

The six farmers who participated in the Land Stewardship Project's monitoring project use many of these practices and have created more natural diversity on their land. Just by converting cropland to pasture they created new habitat for soil microbes, insects, birds, reptiles, amphibians, and small mammals. Species that would have been adversely affected by chemical pesticides and fertilizers used on crops found a more favorable environment in the pastures.

Because of the emphasis on diversity and biological monitoring in holistic management, farmers in the project became advocates of diversity and astute observers of wildlife. A newsletter distributed to monitoring team members contained the following notes in a column called "Farmer Observations":

> Mike saw first red clover blossoms on June 6. Mike saw a hummingbird on clover in his extended rest pad. He suggests that each farmer photograph their rest areas and notice the smell intensified by flowering plants. Ralph saw two baby bobolinks on July 14. He noticed the young are bunching up and may move soon. (Land Stewardship Project 1995)

These farmers are not conventional in any sense of the word. Mike and Jennifer Rupprecht pay meticulous attention to erosion control and species diversity in their pasture, getting excited when they find native prairie species on their land. Ralph Lentz likes to show people the prairie grasses in his pastures and to talk about how he has used managed intensive grazing to improve the stability of stream banks on his land (DeVore 1998). Dave and Florence Minar began working with a local monitoring team, after the original LSP monitoring project concluded, in the area of Sand Creek, the tributary that dumps the most sediment into the Minnesota River. Art Thicke is ecstatic when he talks about the birds he sees while moving cattle—birds that weren't there when those pastures were planted to corn and soybeans (King and DeVore 1999).

The increase of grassland birds wasn't just a phenomenon on Art's farm or on the other five farms in the monitoring project. Other farmers in the Upper Midwest report that they see more grassland birds such as bobolinks (*Dolichonyx oryzivorus*) and dickcissels (*Spiza americana*) since they replaced row crops with grass pastures. The Agriculture Ecosystems Research Project in the agronomy department at the University of Wisconsin has been comparing continuously grazed dairy pastures with rotationally grazed pastures, and preliminary results show that many more birds and more different species use rotational pastures than use continuous pastures (Paine 1996). The increased acres of permanent grass in pasture, combined with conservation reserve land that has been in grass for several years, has created large areas of habitat for game birds also. Additional habitat is created where trees are allowed to grow again along drainages in pastures that were formerly tilled fields.

The farmers actively engaged in the Land Stewardship Project's monitoring project, and many others practicing monitoring as a result of studying holistic management, are protecting or restoring diverse colors and textures in the "warp and woof of the land use fabric." To nurture the diversity of wildlife they have come to appreciate, and the wildlife they have begun to understand as indicators of ecosystem health, these farmers are developing and protecting more habitat niches in wood lots, along roadsides, on orchard and pasture edges, and along streams and ponds. They are leaving areas in their pastures ungrazed during nesting season for grassland birds and removing low areas in fields from cultivation to restore wetlands.

The important point is not that these farmers have become naturalists. The natural habitat they are creating on their land is not because they set out to entice native plant and animal species to reinhabit their farms. Their management decisions and farming practices are turning their farms into a natural habitat for humans, crops, and livestock, *and* wild plants and animals too. Then, as they make the connections between biological diversity, the economic health of the farm, and the quality of their lives, farmers have begun consciously to make decisions to encourage even more biological diversity on their farms. Such farms should be the model for agriculture in the twenty-first century. To make that happen, a large group of constituents are needed who understand the possibilities for farms to be natural habitats and to transform rural landscapes.

Building a Constituency

Aldo Leopold wrote that no government conservation programs with their subsidies for farmers could cause landowners to take good care of the land unless they felt an ethical responsibility for it. The ultimate responsibility for conservation was the farmer's (Leopold 1991). From the latest agricultural census, we can see that less than 2 percent of the U.S. population are farmers (USDA 1997), and not all of them are the family farmers Leopold had in mind but include large-scale farmers managing thousands of acres, often on behalf of investors or on contract with corporations. There aren't enough private landowners on farms to rescue the agricultural landscape from ruin, even if those that exist possess a strong land ethic. We would be foolish to depend upon giant producers and processors such as Tyson, IBP, and Smithfield corporations to exercise a land ethic. Whose responsibility is it then? It is a public responsibility. Good farming produces public goods, and the public must support good farming. Instead of accepting industrial agriculture as a necessary evil and counting on regulations to soften its negative environmental and social consequences, the public (particularly conservationists and environmentalists) should use their dollars and their votes and their influence to bring about agroecological restoration.

If asked whether it is all right to consider agricultural land as an ecological sacrifice area, most conservationists would loudly say no. But without thinking about it, many have acquiesced to the inevitability of farms becoming corporate factories when they have been involved in state or national processes to establish regulations for feedlots. Activist organizations have worked for strong regulations of nonpoint source water pollution and confined animal feeding operations, and their chief opponents have often been farmers, or farm organizations, which has caused them to develop antagonism for farmers. Many haven't had the opportunity to know farmers whose diversified livestock systems operate without need of regulations. If conservationists could get to know farmers who are stewards of the soil, water, and the wild and learn about their management philosophy and the farming practices they use, perhaps they would see possibilities for making basic changes in U.S. agriculture that would restore rural landscapes to greater biological diversity and environmental health.

Dave Palmquist, the interpretative naturalist at southeast Minnesota's Whitewater State Park, the most popular park in Minnesota with about one-third of a million visitors a year, knows a stewardship farm family. He has taken groups of campers 10 miles away from the park to visit the 275-acre farm owned and operated by Mike and Jennifer Rupprecht, one of the six farms in LSP's monitoring project. His reason: "There's an increasing understanding you can't save the world within state parks. The sixty-five little pieces of Minnesota (state parks) aren't going to do it. If you have to go outside your park to tell an important story that relates to the park area, do that." Palmquist believes that visitors are impressed. "It's clear to the visitors that these farmers embrace diversity and see themselves as being part of the bigger environment. The more diversity, the more bobolinks, bluebirds, etcetera, they have on their land, the better they feel. If they can make a living there, maintain a family farm, and be gentler on the environment, that's very exciting for them" (DeVore 1996).

This kind of agroecological restoration is occurring on many farms today, illustrating that farms can be managed to give rural landscapes a mixture of agricultural and natural ecosystems that preserve much of local biodiversity and provide ecosystem services essential to agriculture. We need the heirs to Aldo Leopold's thought and inspiration and those who respect the work of modern ecologists such as David Tilman and naturalists like Dave Palmquist to help society see this vision of the farm as natural habitat and work to turn it into reality.

Conclusion

This vision does not promise that a landscape of such farms will reproduce the ecosystem that existed before white Europeans conquered the land, but neither will it be covered with factories. When farms are factories, they produce commodities and profit for agribusiness and charge external costs to the land and rural communities. When farms are natural habitats for humans, domesticated crops, and livestock, and also for wild plants and animals, they produce food and multiple other benefits for society. And such farms can be the sources for further ecological restoration in the landscape.

No doubt interspersing a variety of uses on farms will mean different problems to overcome than those we now face, both ecologically and economically, because we still have a lot to learn about farming with the wild. Creating farms as natural habitats will require more sophisticated strategies for disease and pest suppression in crops and livestock. It will also require greater emphasis on diversification and resilience and less emphasis on simplification and short-term fixes. These are problems in farming that require ecological solutions.

Farming-system problems can be solved. The perhaps intractable problem is how to influence social evolution so that a land ethic, and not pure utilitarianism, guides land-use decisions. We need all people to look at farming with new eyes, to see the potential of the farm as natural habitat, and to refuse to accept the inevitability of farms becoming rural factories to serve the global economy. We must teach that "the land is one organism."

References

Agricultural Technology and Family Farm Institute (ATFFI). 1996. *Grazing in Dairyland: The Use and Performance of Management Intensive Rotational Grazing among Wisconsin Dairy Farms*. Technical Report no. 5. University of Wisconsin, College of Agriculture, Madison.

Anonymous. 2000. "Farmer Chosen As Next Leopold Center Director." *Leopold Letter* 12(2): 6.

Cambardella, C. A., and E. T. Elliot. 1992. "Particulate Soil Organic Matter Changes across a Grassland Cultivation Sequence." *Soil Science Society of America Journal.*

Community Alliance for Family Farms (CAFF). 2000. *Farmer to Farmer*, May, June.

Council for Agriculture Science and Technology (CAST) 1999. *Benefits of Biodiversity.* Task Force Report no. 133. Ames, Iowa.

Dansingburg, J., and D. Gunnink. 1995. *An Agriculture That Makes Sense: Making Money on Hogs*. Land Stewardship Project, White Bear Lake, Minn.

DeRosier, J. 1998. *My Cover Crop Rotation Program.* Jaime DeRosier, Red Lake Falls, Minn.

DeVore, B. 1996. "An Agrarian Ecological Tour." *The Land Stewardship Letter* 14(4): 2–3.

———. 1998. "The Stream Team." *The Minnesota Volunteer* 61(361): 10–19.

Heffernan, W. 1999. *Report to the Farmers Union: Consolidation in the Food and Agriculture System.* National Farmers Union, Ames, Iowa.

King, T. and DeVore, B. 1999. "Bringing the Land Back to Life." *Sierra* Jan./Feb. 1999: 36–39.

Land Stewardship Project 1995. *Monitoring Project Monthly Newsletter,* June.

Leopold, A. 1966. *A Sand County Almanac with Essays on Conservation from Round River*. Ballantine Books, New York.

———. 1991. "The Farmer as a Conservationist." Pp. 255–265 in *The River of the Mother of God and Other Essays by Aldo Leopold*, edited by B. Callicott and S. Flader. University of Wisconsin Press, Madison.

Minnesota Pollution Control Agency. 1994. *Executive Summary: Minnesota River Assessment Project Report*. Minnesota Pollution Control Agency, St. Paul, Minn.

Paine, L. 1996. "Pasture Songbirds." *Pasture Talk*, May, 8–9.

Rayburn, E. B. 1993. "Potential Ecological and Environmental Effects of Pasture and BGH Technology." Pp. 247–276 in *The Dairy Debate: Consequences of Bovine Growth Hormone and Rotational Grazing Technologies*, edited by W. C. Liebhardt. University of California, Davis.

Savory, A. 1998. *Holistic Resource Management.* Island Press, Washington, D.C.

Soule, J. D., and J. K. Piper. 1991. *Farming in Nature's Image: An Ecological Approach to Agriculture*. Island Press, Washington, D.C.

United States Department of Agriculture (USDA). 1997. *Census of Agriculture*. Vol. 1. United States Department of Agriculture, National Agricultural Statistics Service, Washington, D.C.

Chapter 2

Nature's Backlash

Brian A. DeVore

There's nothing like a creek chock full of dead chubs to make it clear factory farming does not exist in harmony with the local community—natural and otherwise. That became evident to southwest Minnesota farmer Dennis Barta on June 22, 1997, when he stumbled onto the rancid results of a manure spill on Beaver Creek, just a few hundred yards from his house. The manure had originated from a large hog operation some 10 miles upstream. His discovery set off a series of events that eventually led to the apprehension of the guilty party. Fines were imposed and jail time served—a clear example of a lawbreaker caught and punished.

At this writing, the Beaver Creek spill is Minnesota's biggest, documented manure-caused fish kill according to an area fisheries supervisor with the Minnesota Department of Natural Resources who investigated the incident. For a short time, the catastrophe focused the state's attention on the environmental problems associated with factory farms. It didn't hurt that the spill occurred in Renville County, which gained a national reputation during the mid-1990s for the number of factory hog operations that had taken root there. But it's doubtful many people made a connection between the pork chops on their table, the dead clams in Beaver Creek, and the demise of diversified family farming. That's unfortunate, but understandable.

When, as a journalist, I started investigating the Beaver Creek fish kill just a few days after it happened, I focused on the shortcomings of a system that allows large-scale livestock operations to operate with little or no regulatory oversight. As a result, I wrote articles that described the immediate causes of such a catastrophe. I now realize that was only a portion of the story. What happened on that small farm stream during a summer weekend in 1997 has just as much to do with bins full of corn as court proceedings and the way a manure lagoon is engineered.

It would be almost three years after Beaver Creek before a more comprehensive picture of why and how "modern" production systems can go so horribly wrong in farm country would become clear. In those intervening years, I learned not only about what makes a manure spill tick, but also the source of a hypoxic "dead zone" in the Gulf of Mexico, and the demise of a once mighty food production region called the Red River Valley.

The Beaver Creek spill illustrates the immediate causes and shocking results of factory farming going awry and the impacts on a local community. The Gulf of Mexico's hypoxia problem shows what happens when numerous "Beaver Creeks" gang up to form one big regional problem far downstream. And to really get at the heart of the problem, I had to go to one of the nation's richest farming regions. There I saw how the same farming system that was killing fish in Beaver Creek and the Gulf of Mexico was also destroying farm town Main Streets. The real problem is not crimes that make the county sheriff's report. The problem is farming practices that no one goes to jail for. In fact, such techniques are often encouraged and rewarded. Agricultural science has created tools to bulldoze nature out of the way and replace some ecosystem processes with technology, but it doesn't always overcome or outwit nature; instead it sometimes creates bigger problems for society.

What I learned is that these seemingly isolated problems associated with industrial agriculture are in reality intricately intertwined. The true causes of fish kills on farm creeks extends much further upstream than the closest manure lagoon. Discover why we have fewer bobolinks in central Iowa, contaminated well water in southeast Nebraska, or shuttered main streets in Illinois, and other answers will follow. These answers illustrate the need for a new vision of agriculture, as is proposed in this book, to put farming systems back into ecosystems.

Anatomy of a Manure Spill

Dennis Barta speaks in a flat, laconic tone that gives the same emotional heft to everything he talks about, from the weather to what it's like to see firsthand a major ecological catastrophe. One has to listen carefully when he describes what happened on a Saturday evening in June 1997; otherwise, the point of the story may be missed: he witnessed a prime example of concentrated livestock production going haywire.

It all started when Barta and his son Nathan went to Beaver Creek, which winds its way through their farm, for a few hours of fishing. Barta grew up on this farm and has been raising crops and livestock on it since the early 1970s. He has been catching chubs out of that creek since he can remember. Barta's father also fished the waterway, which is about a dozen feet across and just deep enough to swamp a chore boot. But the fish weren't biting that particular summer evening.

"I think they were too busy fighting for air," Barta told me when I visited Renville County a few weeks after the incident. "Then the creek turned white with the bellies of dead fish. I knew something was wrong, but I didn't know what."

What the farmer and his neighbors figured out by dawn the next day was that as much as 100,000 gallons of manure had washed into the creek the previous Thursday evening, making the waterway a biological dead zone. Dennis and his son had gotten a firsthand glimpse at the climax of the kill. The manure was produced by a large-scale confinement operation that was raising pigs for one of the biggest pork production companies in the nation. The official reason given for the spill was a "faulty timer" on a washing mechanism. According to the Minnesota Pollution Control Agency, this caused a water pump not to shut off, sending a pureed mix of manure and urine overflowing from a pit underneath one of the facility's buildings. It swamped the pigs inside but quickly escaped out the building's doors, saving the animals from drowning in their own waste. The river of feces then traveled about 100 feet overland to an open pipe in a field. The pipe, which was part of a tile drainage system used to keep standing water off the field, provided the manure a one-way ride to Beaver Creek.

When the manure hit the water, bacteria started feeding on it in a supercharged frenzy, consuming oxygen at an extraordinary rate as they respired. The fish—more than 690,000 of them according to the Minnesota Department of Natural Resources' estimate—literally suffocated along 18.6 miles of the stream. In addition, ammonia present in the waste made the water toxic. Sixteen species of fish were affected by the spill, but that doesn't take into account the innumerable insects and invertebrates that were killed. Fisheries biologists I talked to wonder if long-lived species like clams will ever return to the stream in any significant numbers. Several local residents told me that at the peak of the spill they saw fish clustered around places where fresh water was trickling into the creek, desperate for any new source of oxygen. Even more troubling was that crayfish, tough bottom-dwelling crustaceans that are seldom seen on dry land, were literally clawing their way up the banks to get away from the water as rafts of manure floated by.

More than wild residents of Renville County were affected by the Beaver Creek spill. Before it joins the Minnesota River (which eventually flows into the Mississippi River), the stream flows through a county park that serves as a popular fishing and swimming spot. The creek also runs through a pasture where Richard and Mary Nina Serbus graze dairy cows. The weekend of the spill, "bogs" of manure came floating through their land, Richard Serbus said. A water sample taken from Beaver Creek where it runs through the Serbus property two days after the spill showed a fecal coliform level that was 3,800 times higher than the Safe Drinking Water Act's maximum contaminant level (MVTL 1997). The cows, which often drink from the creek, got high fevers, and many eventually aborted their calves. Milk had to be dumped while the sick cows underwent treatment with antibiotics. The Serbus family also suffered bouts of diarrhea after the spill. After their well water tested high for contaminants like nitrates, a public health nurse recommended they not drink it. These days, the Serbuses do drink the water from their well, but not until it passes through a reverse osmosis machine that sits on their kitchen counter. They also stay out of the creek.

"The kids used to make rafts and float around or get inner tubes and float down the creek," Mary Nina told me, glancing down hill toward the creek. "We wouldn't dare do that today."

The Beaver Creek Manure Spill of 1997 has it all, even intrigue and a true-to-life "bad guy." According to the Minnesota Pollution Control Agency, the man who owned the guilty operation, Roger Kingstrom, did not report the spill until three days after he discovered it. After a delay of several months in which various state and local agencies grappled with who was responsible for enforcement, Kingstrom was jailed for thirty days and fined. But after that he went on farming and running a manure hauling business (he was given work release while in jail so he could manage his hog operation during the day). And the firm Kingstrom was raising the pigs for, Christensen Farms, continued to rise in the ranks as one of the ten largest hog companies in the nation (Freese 2000).

As large livestock confinement systems and their multi-million gallon manure storage facilities proliferate throughout the countryside, the Beaver Creeks of the world are becoming commonplace. That's not to say manure from farms of all sizes hasn't found its way into our nation's waterways in different proportions in the past. But for the first time in history, agricultural waste accidents are taking on the same size and import as, say, oil spills. Factory farming has earned significant clout in at least one area: its industrial-sized ability to produce pollution. One large animal confinement operation can produce as much waste as a small city, and the United States generates 1.4 billion tons of animal manure every year—130 times more than the annual production of human waste. No wonder animal waste is the largest contributor to pollution in 60 percent of the rivers and streams classified as "impaired" by the Environmental Protection Agency (Wright 1999).

Into Thin Air

Like Dennis Barta, Donald Lirette has a story about dead fish gasping for air, but on a much wider scale. In the early 1980s, he was working as a shrimper in the Gulf of Mexico, which meant his work days often consisted of cruising around the region in a four rigger boat, using "try nets" to figure out if he should stop long enough to drop bigger nets into the water. On one outing, Lirette quickly became aware that the try nets were dredging up nothing but bad news.

"As soon as that try net would hit the deck, it would smell decomposed," he told me in his high-octane Cajun cadence. He was speaking over the telephone from his home in Louisiana's Terrebonne Parish, one of the most fishing-dependent areas in the region. "Even the hermit crabs were dead, and nothing kills hermit crabs."

Lirette revved up the powerful diesel engines on his trawler and tested the waters further out. But it soon became clear a few gallons of fuel wouldn't carry him past this problem.

"I thought it was caused by some local discharge and then I just kept traveling and traveling and then I realized this wasn't local."

In fact, it was the harbinger of a problem that has linked one of the most productive agricultural regions in the world with one of its most vital fisheries. Within a few short years, the "hypoxic" (low oxygen) zone forming in Lirette's backyard has become one of the biggest environmental issues of the decade. And the cause is excessive fertilizer and manure runoff from Midwestern farms. The Gulf "dead zone" issue has become the ultimate example of how what is done on a farm has impacts well beyond its fenceline.

If a manure spill kills like a car crash, the Gulf's hypoxic zone is cancer, slower but just as ultimately deadly—and with a much wider area of impact. Key research reports done by a federally mandated Nutrient Task Force and the Council for Agricultural Science and Technology have concluded that excessive agricultural nutrient runoff is the cause of the Gulf's woes (NOAA 1999, CAST 1999). The nutrient scientists are pointing a particularly big finger at nitrogen, which is a linchpin fertilizer for Midwestern row-crop production. The nitrogen cycle makes the biological world go 'round and since World War II (thanks to bomb-making technology) agriculture has controlled this key to plant growth. The result has been unleashing massive amounts of a nutrient upon the environment that in the past was only made available through certain specialized bacteria or lightning.

In fact, nutrients such as nitrogen to some extent are responsible for making the Gulf such a rich fishery. But when too much of a good thing hits that salty water, it sets off a fish-killing chain reaction that is similar to what happened on Beaver Creek: a super-growth of phytoplankton results from the influx of nutrients. As the phytoplankton dies, it consumes oxygen, particularly close to the bottom. That produces a zone so low in oxygen that the fish flee—or die.

Such low levels of oxygen affect the fishing business in two ways. For one thing, it requires shrimpers and others to travel farther and farther to fill their nets. But perhaps even more importantly, the hypoxic zone serves as a biological force field that blocks fish from traveling between spawning grounds and other parts of the Gulf. Even though it represents only about 1 percent of the Gulf's total area, this necklace of sick water (it's been as much as 300 miles wide) is in a strategic location when it comes to the region's ecological health. At its peak, the zone stretches from the mouth of the Mississippi past where the Atchafalaya River enters the Gulf. Between those two waterways is one of the richest aquatic systems in the world (CAST 1999).

Fisheries in areas like the Black Sea and France's Sommons Bay have been devastated as a result of hypoxia. The Gulf may not be far behind. By the early 1980s, shrimp catches there were dropping dramatically in areas where bottom waters were hypoxic. In the hardest hit areas, a boat hauling a 40-foot net for six hours might not catch a single shrimp (CAST 1999). Shrimpers can often fill out their quotas by going to the edge of the zone, where escaping aquatic life is heavily concentrated, but people like Lirette worry about the long-term future of the Gulf's sport and commercial fisheries, which together produce $2.8 billion in economic activity annually.

"A few years ago a friend called and said, 'Hey Donald, we are dipping shrimp right off the beach with dip nets,'" he recalled. "They were probably going out to spawn and got turned around by the dead zone, and got cornered next to the beach. We call that a jubilee. It was good for my friend, but it was probably not good for the shrimp."

And it's getting worse for the shrimp. Between 1993 and 1996, the Gulf's hypoxic zone averaged 4,000 square miles. By 2000, the five-year running average was 5,454 square miles (EPA 2001).

Part of the problem can be traced to the massive amounts of nitrogen present throughout the watershed year in and year out, no matter how much water is flowing into the Gulf. In fact, the amount of nitrogen entering the Gulf has more than doubled over the past four decades. About half of that nitrogen is from commercial fertilizer and 15 percent is from livestock manure. Most of it is coming from the Corn Belt: Iowa, Illinois, Indiana, Ohio, and southern Minnesota. In fact, the upper Mississippi basin (above the Missouri River) comprises about 15 percent of the drainage area of the entire watershed but contributes more than half of the nitrogen discharged to the Gulf (CAST 1999).

But let's not focus too much on nitrogen as the cause of the Gulf's environmental woes. Rather, we need to consider a deeper question: What kind of land-use system allows a wayward nutrient to do so much damage so far downstream?

The Numbers Game

On a recent late summer day, I drove through extreme southwest Minnesota, a landscape dominated by succulent corn stalks and rows of soybeans stretching across fields like fat ropes of green velvet. Near the town of Lamberton, I pulled into the Southwest Minnesota Research and Outreach Center. There, on one wall of the agricultural experiment station's large meeting room, was the story of a region becoming one of the least biologically diverse areas on earth within a span of a few decades. Fifty sheets of blue paper used computer-generated charts and graphs to show a two-crop monoculture of corn and soybeans methodically replacing a system that, at the turn of the century, included wheat, oats, barley, rye, alfalfa, and pasture. (For a similar graph, see chapter 10, where Laura Jackson analyzes similar statistics for Iowa.)

By the time I finished looking at this wall of agronomic history, it wasn't a shock to learn that 91 percent of the cropped acreage in a nine-county area of Minnesota is now planted to either corn or soybeans. Sixty-seven percent of the region's total land area is growing one of those crops. This wall of statistics also tells the story of how diverse cropping systems stopped being a key farming tool and evolved into a modern farming liability (Getting et al. 1999).

Before the advent of agrochemicals, farmers needed livestock and diverse cropping systems to return nutrients back to the land and control pests. For example, a farmer would raise cattle on hay, oats, corn, and pasture. The manure from those cattle went back to the land that produced the feed, and the cycle started over again. After World War II, it was assumed by most farmers that synthetic fertilizers and chemical pesticides made the fertility-building, pest-killing abilities of diverse cropping rotations superfluous.

At the same time, the government was paying farmers to plant corn and wheat, not alfalfa, rye, and pasture. As the diversity declined, and tractors became affordable enough to be purchased by individual farmers, it became easier to farm more acres. As such equipment became larger and more specialized, bigger, uniform fields made more sense, and vice versa.

Livestock were increasingly taken off cropping operations and raised in concentrated numbers by factory farm operations as farms became more focused on one or two commodities. Animals in these facilities eat high-energy feed derived mostly from corn and soybeans that may be raised in the next township, the next state, or overseas. Eighty percent of U.S. corn is made into animal feed for both domestic and international use (NCGA 2000). Such a system produces vast quantities of liquid manure. Instead of being a sustainable way to return nutrients to the fields that produce feed, manure has become a waste product. The cycle that circulated nutrients from the land, through animals, and back to the land again has become a one-way trip that overwhelms the ecosystem.

Such changes have brought about a tremendous disruption of the nutrient cycle, and during the latter half of the twentieth century man-made nitrogen became the cornerstone of maintaining fertility in cropped soils. A photo from a 1953 high school chemistry textbook illustrates why. It shows a stand of corn that's short, spindly, and pale in color. Right next to it are two rows of corn that are tall, green, and succulent. The sickly looking stand received no nitrogen fertilizer and produced 24 bushels per acre, according to the photo caption. The green giants got nitrogen and produced 110 bushels per acre (Hogg et al. 1953).

It's no accident the Upper Midwest accounts for more than 80 percent of the nation's corn crop and better than half of the amount of commercial nitrogen fertilizer applied (CAST 1999). The cost of nitrogen fertilizer is still relatively cheap (it makes up less than 10 percent of the total input costs of producing a bushel of corn) and farmers often apply more than the crop needs because of its tendency to leach out of the soil. Farmers have dramatically reduced nitrogen application rates in recent years. In 1985, the average Iowa crop farmer applied 145 pounds of nitrogen on each acre of corn raised. By 1996, that rate had dropped to 125 pounds (Baker et al. 1996). When spread over millions of acres, that adds up. But agronomists say in general too much nitrogen is still being used on our nation's croplands. Soil scientists and economists have told me that farmers often opt to stick with higher rates because they consider it "cheap insurance."

Don't tell Donald Lirette that. If one were to lay those southwest Minnesota cropping history charts over a graph showing how much nitrogen enters the Gulf every year, the parallel rising lines would be hard to dismiss as mere coincidence. As more corn and beans were planted, more livestock became concentrated on fewer farms, more pasture, hay and small grains went by the wayside, and nitrogen levels in the water went up.

The Blame Game

Whom do we blame when industrial agriculture does so much wrong to the land and its people? In the case of Beaver Creek, it would be easy to point the finger at Roger Kingstrom, the farmer who owned the facility, or Christensen Farms, who owned the pigs. Or how about the teenaged employee who was on duty at the time? The faulty timing mechanism? Easy blame begets easy solutions: just ban Kingstrom from farming; fine Christensen Farms so they will think twice about how and by whom they have hogs raised; make it illegal for anyone under eighteen to manage a livestock factory alone (Renville County officials discussed such an option); place an earthen berm around all confinement facilities to contain overland flows of runaway manure; or require that every washer timing mechanism have a second timing mechanism backing it up.

Whose fault is it we have hypoxia in the Gulf? Corn farmers? The scientist who first figured out how to cook up nitrogen fertilizer in the laboratory? The people who dig trenches and lay drainage pipe to siphon off more surface water (and whatever it contains)? When the Gulf hypoxia issue became big news in the mid-1990s, agronomists and agricultural engineers held endless discussions on how field drainage systems could be designed differently to divert nutrient runoff, or how satellite technology could be used to place fertilizers more precisely.

All of these measures may help in isolated cases, but they will do little to solve the overall problem of relying on a simplified agricultural system that must be propped up by lots of energy, chemicals, and water.

Chinks in the Armor

Those charts I saw at the southwest Minnesota experiment station were generated by Paul Porter, a University of Minnesota agronomist who's been working on ways to promote more cropping diversity in farm country. On the day I visited the station, he took a break from his field work to talk about a "frustration" he shares with many other researchers: the soil's ability to put up with an abusive cropping system.

"We can show there's been a drop in organic matter all across the Midwest and yet at the same time look at our yields and consider the trend," he said. "It's not obvious that we are plateauing yet. This soil is just too good."

That's a key point. It's long been argued by agribusiness that farmers would never do anything that harms the environment. After all, if they were truly degrading the land, their crop yields would suffer. Economic self-interest is the best stewardship incentive, goes this argument. But a combination of rich soils and chemical inputs has provided the system with a nice disguise, allowing it to hide its innate unsustainability. When the traditional nutrient cycle was disrupted and organic matter was lost through erosion, man-made nitrogen fertilizer plugged the hole that was created. When diverse plant systems were eliminated, pesticides helped make sure bugs and weeds didn't take too much advantage of the situation.

However, there are signs the system is running on empty. Although record yields of corn are still common, there is some concern that the peaks and crashes of crop yields are becoming more extreme (University of Minnesota 1998). In addition, a cyst nematode began attacking Midwestern soybeans in the 1990s, while in Illinois, farmers have made the troubling discovery that certain western corn rootworm beetles are able to survive a season in a field planted to soybeans. That's not supposed to happen: as their name implies, these beetles normally die when fed anything but corn. In fact, soybeans rotated with corn every other year has traditionally been used as a way of breaking up the breeding cycle of this pest. It appears that strategy is failing (Thomas 1998).

Porter estimates that as a percentage of total direct expenses, the seed and crop chemical input costs for corn and soybeans increased from about 20 percent to about 25 percent between 1999 and 2000. One important reason for that higher cost is increased pressure from insects, weeds, and diseases associated with a greater reliance on a simplified cropping system, according to Porter.

Homegrown Problems

Such wrinkles in modern industrial farming tend to bring the problems of this system closer to home. Farmers are increasingly seeing firsthand the impacts of oversimplifying the ecosystem in places like the Red River Valley of the North. For more than a century, replacing complicated biodiversity with a handful of crops made the Red River Valley the stuff of agricultural lore. Soils deep enough to bury a man upright helped make this 300-mile flattened trough of former tallgrass prairie a world class wheat and barley producer. But in the mid-1990s, this simplified system proved to be the perfect environment for a fungus called *Fusarium graminearum* (wheat scab). Year after year of the same crops planted on contiguous sections of farmland have made it easy for the fungus to survive and spread, particularly under wet conditions. That's no surprise to University of Minnesota ecologist David Tilman, who has studied the effects of replacing biodiversity with monoculture.

"When we set out a huge part of the landscape to a single plant species, the pathogens have it easy. If we want agriculture to survive, we have to outsmart pathogens," he told me. "We're not outsmarting them right now. We're playing the dumb game."

It's proving to be an expensive game as well. Small grains growers in the region lost $4.2 billion worth of income between 1992 and 1998, mostly because of the scab (McMullen et al. 1997, University of Minnesota 1998). Farmers desperate to save their crops sprayed more fungicides than ever before, releasing some of the most toxic pesticides known into the environment.

About one-fifth of the Valley's farmers went out of business in 1997 alone. On the Minnesota side of the Red River of the North, wheat plantings were down almost 30 percent in 1998. The region—it covers parts of Canada, North Dakota, South Dakota, and Minnesota—is no longer one of the most productive places on the planet for small grains (University of Minnesota 1998). In fact, some agronomists are beginning to wonder if cereal grain crops will have *any* role to play in this region's future. "The ideal situation is that wheat and barley don't disappear from this area," said Jochum Wiersma, a small grains specialist at the Northwest Minnesota Agriculture Experiment Station. He told me this one September morning as he thumbed through pages of increasingly dreary statistics on the region's crop-planting history. It wasn't exactly a ringing endorsement of the Valley's future as a small grains producer.

The Red River Valley entered the twentieth century a prime example of the good that can come from specializing crop production. But it departed the millennium with a different legacy, as an example of the bad that results from destroying all that biodiversity.

Personal Choices—Wider Changes

Individual farmers who make changes in their production practices are often prompted by an immediate threat to their family's well-being: a high nitrate test in the well water, respiratory problems caused by raising livestock in confinement, and so forth. Indeed, as a result of the Red River Valley's agronomic collapse, desperate farmers are showing more interest than ever in alternative systems of production that utilize diversified rotations. Even bankers in the region are suggesting that their farmer clients check out organic agriculture, say extension educators I've interviewed. In the late 1990s, the Southwest Minnesota Research and Outreach Center started sponsoring field days that focused entirely on sustainable cropping systems. They regularly drew more than 100 farmers, which would be considered a big turnout even for a field day on "conventional" cropping systems.

In order to spawn widespread changes that don't just reduce nitrate leaching here, save a few fish there, or keep a farm family financially viable someplace else, initiatives must be taken that reach across boundaries—political, philosophical, and economic—and help people recognize some common enemies. There are some good examples of that thinking already being put into action. Dan Specht, a northeast Iowa farmer, has been to the Gulf of Mexico as the invitation of an environmental group. While there, he saw how a leaky agricultural nutrient cycle was causing contaminated water in his home county *and* dead shrimp in Louisiana.

"When you solve the problems of the Louisiana fishermen, you're also solving drinking water problems for millions of people in the Midwest," said Specht.

An even more comprehensive example of this kind of big picture thinking dates back to the spring of 1998. That's when a sixty-member task force consisting of Minnesota farmers, as well as representatives of agribusiness, public agencies, and nonprofit organizations concerned about everything from the environment to rural social justice, concluded that the crisis in the Red River Valley was a prime example of technological fixes such as chemicals failing to make up for a lack of biological and genetic diversity (University of Minnesota 1998). It was an astonishing conclusion not least because of the diversity of people who came up with it. But, as this task force has proved, sometimes desperate times spawn a collective recognition that there is more at stake than declining corn yields or belly-up bullheads.

"It isn't just an environmental tragedy that's developing," Don Wyse, a University of Minnesota weed scientist who co-chaired the task force, told me after the report came out. "It's also an economic, family, quality-of-life thing as well. The resilience is being lost in terms of the environment, but it's also being lost in terms of the people. It's fragile all the way through."

References

Baker, J. L., S. W. Melvin, and P. A. Lawlor. 1996. *Agricultural Water Quality Improvement: Nitrogen Management Research Results*. December. Iowa State University, Department of Agricultural and Biosystems Engineering, Ames, Iowa.

Council for Agriculture Science and Technology (CAST). 1999. *Gulf of Mexico Hypoxia: Land and Sea Interactions*. Council for Agricultural Science and Technology, Ames, Iowa.

Freese, B. 2000. "Pork Powerhouses 2000." *Successful Farming* 98(11):18–24.

Getting, J., P. Porter, and M. Werner. 1999. *Trends in Minnesota Agriculture*. September 23. University of Minnesota, Southwest Minnesota Research and Outreach Center, Lamberton, Minn. Available online at http://www.rrcnet.org/~swes/TIMNA/HOMEPAGE/contents.htm.

Hogg, J. C., O. E. Alley, and C. L. Bickel. 1953. *Chemistry: A Course for High Schools*. D. Van Nostrand Co., New York.

McMullen, M., R. Jones, and D. Gallenberg. 1997. "Scab of Wheat and Barley: A Reemerging Disease of Devastating Impact." *Plant Disease* 81:1340–1347.

MVTL Laboratories (MVTL). 1997. Water Sample Results from Section 6, Bird Island Township, Co. Rd. 14 Bridge. Work Order Number 6724. June 26. MVTL Laboratories, New Ulm, Minn.

National Corn Growers Association (NCGA). 2000. *Corn Curriculum*. National Corn Growers Association, St. Louis, Mo.

National Oceanic and Atmospheric Administration (NOAA). 1999. *Hypoxia in the Gulf of Mexico*. National Oceanic and Atmospheric Administration, Washington, D.C.

Thomas, J. 1998. "With Change in Habit Beetles Shake Up Corn Country." *The New York Times*, October 20, sec. F, pp.1–2.

University of Minnesota Extension Service. 1998. "Farm Income Plummets in Southwestern and Northwestern Minnesota." *News and Information*, April 29. University of Minnesota Extension Service, St. Paul, Minn.

U.S. Environmental Protection Agency (EPA). 2001. *Action Plan for Reducing, Mitigating, and Controlling Hypoxia in the Northern Gulf of Mexico*. January. U.S. Environmental Protection Agency, Mississippi River/Gulf of Mexico Watershed Nutrient Task Force, Washington, D.C.

Wright, A. G. 1999. "A Foul Mess: EPA Takes Aim at Factory Farms, the No. 1 Water Polluter in the U.S." *Engineering News-Record* 243:14–26.

Chapter 3

The Farm, the Nature Preserve, and the Conservation Biologist

Laura L. Jackson

We sit in a circle of hay bales on a hill overlooking a pothole marsh, a fall-gathered flock of gleaming white pelicans just visible below us. I've been invited to a dozen or so of these farm tours over the past several years to be the official "ecology expert," but I still don't know what to say. The farm is beautiful—several neat acres of vegetables surrounded by windbreaks of hybrid poplar, kept-up outbuildings, a strip along the farm lane planted to brome, big bluestem, and shrubs, and a lovely view of the state-owned marsh from the farmhouse kitchen. The landowners rent out most of the farm ground surrounding their farmstead to a neighbor, and it is planted to corn and soybeans. The husband has an off-farm job, and the wife and kids sell vegetables and guinea hens by subscription to people in the small towns nearby.

A part of me is very comfortable. I spent my teens on an acreage like this, helping my mother in the garden and tending livestock for our family's consumption. But the landowners have asked me to talk about how this farm could contribute to the restoration of greater biological diversity in the Midwestern landscape, and in that role I am ill at ease. Shall I tell them to cut down the hybrid poplars, sever their relationship with the neighbor who rents ground from them and plant it to prairie, rip out the brome and shrubs, and sow prairie wildflower seeds along the farm lane? Sell the guineas and turn loose prairie chickens? Build birdfeeders and a cabin for ecotourists?

The midwestern United States is blessed, or perhaps cursed, by deep black soils, ample summer rainfall, and gentle topography. Once a vast region of prairies, savannas, river-hugging forests and wetlands, it is now managed for agricultural production. The intensity of this management increases yearly, as technological, social, and economic forces eat away at the remnants of native vegetation and wildlife habitat on farms. "Nature" in the most common sense of the word is relegated to small protected areas, usually on marginal soils too steep, rocky, or boggy to farm.

For many conservation biologists and card-carrying members of the national environmental organizations, this has been an acceptable compromise. Plain farming country in North America is a kind of ecological sacrifice zone with a conservation profile as low as its perceived cultural status. The sacrifice is more than metaphorical. Some agronomists and conservationist biologists, seconded by spokesmen for Monsanto, have proposed that farming practices be intensified even further in northern temperate zones by the use of increased fertilizers, pesticides, and biotechnology so that the tropical rainforests might be spared (Horsch and Fraley 1998, Avery 1995, Waggoner 1994, Huston 1993). Indeed, parts of the Amazon River basin are currently being converted from forest to soybeans, and rivers dammed for barge transportation, in active competition with North America. If we restrict soybean acres here and drive up soybean prices, it is likely that more Amazonian forest will burn.

In the last century of farming, Midwesterners have hastened to purchase and protect the last remnants of primary (i.e., never clearcut or plowed) forests and grasslands. Perhaps no collection of nature lovers is more protective of its preserves than the prairie enthusiasts of the Upper Midwest. Meanwhile, the science of conservation biology has paid a great deal of attention to the protection of populations of rare species, particularly those whose numbers are dwindling in fragmented and isolated habitats. Despite this, relatively little attention has been paid to the changing nature of farming surrounding the archipelago of nature preserves in farming country. Once sprinkled with pastures, hay meadows, grassy fencerows, hedgerows, wetlands, and riparian areas, the working farm landscape was adequate to serve as stepping stones between preserves. Now, the matrix of privately owned farmland has become a forbidding ocean separating distant nature preserves and threatening their existence.

It is the goal of this book to raise the profile of conservation on agricultural lands and transform that ocean. In this chapter, I will review what is wrong with depending on nature preserves alone to protect nature. Then I will look at the question of what prevents conservation biologists and others from dealing with the "elephant in the living room"—biological diversity conservation on privately held agricultural lands.

The Problem with Protection

Nature preserves in agricultural areas are surrounded by increasingly hostile territory, and evidence is mounting that a smattering of small protected areas throughout a region—any region—will not be sufficient to protect biological diversity much less protect the water or the air. Consider Daubendiek Prairie, a 4-acre gem of tallgrass prairie in northern Iowa and a microcosm of many of the problems facing small fragments of natural habitat in farm country. Initially considered too wet to plow, it was bought in the 1970s by a nurseryman named John Daubendiek and dubbed "Johnny Walnutseed," who planted black walnut saplings directly into the prairie sod. This move apparently preserved it from later, more aggressive rounds of tiling (subsurface drainage) that could have brought it into production, because only the saplings on the perimeter of the prairie took hold, while those in the boggy interior failed to thrive. Daubendiek's prairie, studded with these dead walnut saplings, is bordered on two sides by corn and soybean fields, on one side by an abandoned field last farmed fifty years ago, and on the fourth by a gravel road.

My husband and I lease it from the landowner to prevent it from becoming a corn field, but the prairie is bombarded by things that move with impunity across legal boundaries. Pollen from corn drifts in on the wind, and if impregnated with the Bt toxin through genetic engineering the pollen can hurt the butterfly caterpillars nibbling on prairie plants. Silt, weed seeds, and fertilizer wash in from the fields above, planting a stand of giant ragweed three meters tall. An underground tile line surfaces in the old field, draining excess water and farm chemicals from the fields above. Kentucky bluegrass creeps in from the edges, and bird droppings carry the seeds of Tartarian honeysuckle to the row of burr oak trees along the farm lane. On the sides nearest the corn, one finds empty bags of insecticide, piles of fieldstones, and a diesel fuel tank. The landlord recently pulled onto the edge of the property, found he couldn't back out, and had to drive a big loop, mashing plants and creating muddy ruts along the way.

Although the planted walnuts never amounted to much in the boggy interior, the native quaking aspens, sumac, hazelnuts, gray dogwood, raspberries, and other woody plants are thriving, crowding out the grasses in the absence of frequent wildfire. With the trees have come brown-headed cowbirds, a native species that lays eggs in other birds' nests. When the cowbird eggs hatch, they push out the other nestlings and take all the food their foster parents bring home.

Meanwhile, small protected areas like Daubendiek Prairie may be nothing but a final resting place for the "living dead"—organisms that cling to life but fail to reproduce due to lack of pollinators, mates, or other reproductive requirements. Each May I find about thirty clumps of yellow ladyslipper orchid and only two clumps of the endangered small white ladyslipper orchid. Like a gambler starting the night with just a few dollars, these populations are vulnerable to a simple run of bad luck. A few lousy hands in succession, such as bad weather, disease, or attacks by orchid poachers, could extinguish these small populations. Close inbreeding and loss of genetic diversity may eventually weaken them further. And once lost from this isolated spot, they will likely not come back. The preserves are too far apart to allow all but the most mobile animals, like birds, to recolonize them.

Daubendiek Prairie shares these problems with all small- and middle-sized nature areas in the agricultural Midwest. We can compensate for the absence of wildfire with carefully managed prescribed burns. Boundaries can be policed for exotic species control and unauthorized vehicle traffic. Genetic erosion can be prevented by artificial pollen transfer among populations. All this takes time, money, and people.

The largest nature preserves can probably manage their way out of most of the outside influences mentioned above, but a strategy of reliance on large nature preserves has many flaws. First, large preserves of several thousand hectares or more are exceedingly rare in agricultural landscapes—perhaps one or two per state. And even very large (by Midwestern standards) nature preserves face problems. Studies on the 8,600-acre (3,487-hectare) Konza Prairie in Kansas have shown that bison had a strong influence on the plant community, but most big preserves can't afford to manage cattle, much less bison, for conservation purposes. Even the very largest nature preserves (including the Konza) are far too small for large carnivores (wolves, black bears, mountain lions), that used to form the peak of the food pyramid. It has been hypothesized that the absence of large carnivores has created an overabundance of "mesopredators"—skunks, raccoons, crows, foxes, possums that in turn prey heavily on smaller animals such as birds (Soulé et al. 1988).

Because running water knows no preserve boundaries, fish and other aquatic species are vulnerable to the everyday perils of heavy sedimentation from soil erosion as well as to lethal manure spills, nutrient runoff, and pesticide contamination. Some pesticides vaporize on hot days and are carried far off the field as fumes, killing sensitive plants and trees. In both cases, the effect of being surrounded by agriculture carries deep into preserves, large and small.

Anhydrous ammonia application (the nitrogen source of choice for corn production) and manure lagoons used in factory-style livestock production release gaseous ammonia, which wafts up into the atmosphere and returns to earth as nitrogen-fertilized rainfall. This puts many species of plants that are adapted to survive under *low* nitrogen conditions at a competitive disadvantage, and they disappear from the ecosystem (Berendse et al. 1993, Tilman and Wedin 1991). There is new evidence that nitrogen-enhanced rainfall is changing soil microbial communities as well. Edgerton-Warburton and Allen (2000) found stark changes in the species composition of arbuscular mycorrhizae (soil fungi that form intimate associations with plant roots) along a nitrogen deposition gradient that was created by auto emissions in southern California.

Even large preserves are unable to influence the hydrology above them in the watershed. Next to climate change, hydrology holds the greatest veto power on our ability to maintain or restore biological diversity. The hydrological cycle has been radically modified in virtually every corner of the agricultural Midwest simply due to the replacement of the sponge-like prairies with plowed ground. For instance, the 5,500-acre (2,226-hectare) Neal Smith National Wildlife Refuge in south-central Iowa cannot reconstruct its riparian wetlands because water no longer flows into the refuge slowly and evenly but in short violent bursts. Aldo Leopold was particularly interested in this phenomenon and made reference to the changed nature of creeks in *A Sand County Almanac* (1966): "In the creek bottom pasture, flood trash is lodged high in the bushes. The creek banks are raw; chunks of Illinois have sloughed off and moved seaward. Patches of giant ragweed mark where freshets have thrown down the silt they could not carry" (127).

Reliance on large terrestrial preserves will not save our rivers and estuaries. All of the large rivers have been radically altered to move agricultural products on barges and protect valley farmland. On the Missouri River, abandoned channels of the ever-restless river called "oxbows" were once popular recreation areas, but they began to dry up in the 1970s. Dams upstream had caused the river to cut a lower channel for itself and the groundwater that kept the lakes filled dropped. Channelization of the Missouri meant that farmers could plant to the water's edge and that industries like Archer Daniels Midland could safely locate their manufacturing plants in the floodplain. Ultimately, nearly three-quarters of a million acres (1,150 square miles) of continuous river valley habitats—floodplain forests, wetlands, sandbars, and backwaters—were converted to farms and paved areas (Schneiders 1999). The pallid sturgeon (*Scaphirhynchus albus*), a fish whose ancestors roamed the large inland rivers of North America 70 million years ago, is now endangered due to lack of appropriate spawning sites. According to a February 15, 2000, CNN news report, the catastrophic floods of 1993, which destroyed many water control structures on the Missouri, must have offered the pallid sturgeon a rare opportunity to reproduce. We are now starting to catch fish that spawned in 1993, essentially the only pallid sturgeons in the river that are under thirty years old.

The upper Mississippi River suffers from sedimentation of shallow backwaters due to upland soil erosion rates far greater than those of the prairies. The system of locks and dams along the river allowing barge transport of agricultural chemicals and commodities further exacerbates the problem. The once abundant wetlands fringing the river channel used to feed migratory waterfowl using this important flyway. Now, according to U.S. Fish and Wildlife Service staffer Cathy Henry, their agency has to manage a series of artificial wetlands they call "moist soil units" up and down the river to help feed the birds. In the United States, 28 percent of all amphibians, 34 percent of fishes,

65 percent of crayfishes, and 73 percent of all freshwater mussels are ranked extinct, imperiled, or rare by the Natural Heritage Network of the Nature Conservancy (Allan and Flecker 1993). Given the fact that everything in the water must deal with the cumulative impacts of land use and hydrological modification, these numbers are hardly surprising.

Croplands, pastures, and rangeland make up about half of the surface area of the United States and are managed by just 2 percent of the U.S. population (NRCS 1996). What happens on these lands matters greatly to the 98 percent of people who are not managing this land, making agriculture a public good and not just a business. Activities on private agricultural lands affect everything from drinking water and invertebrate communities in the local creeks, to the fish nurseries and shellfish beds in major estuaries, to coral reefs. Even the largest, well-managed nature preserves are at the mercy of larger-scale forces that originate on private land.

In summary, whether in agricultural areas or other managed landscapes, nature preserves alone cannot preserve all nature. Plants and smallish- to medium-sized terrestrial animals are probably adequately protected by the largest well-managed preserves, but large terrestrial mammals and aquatic organisms of all varieties are still highly vulnerable. Some kinds of biological diversity, such as genetic variation within species and multiway interactions among plants, soil, insects, and microorganisms, are largely invisible and therefore unmanageable, as are many large-scale ecological processes.

And what happens to people in a society where nature is relegated to a few large, nature preserves well managed by trained experts? There is something to be said for nature enjoyed and, yes, managed, by a wide variety of people across the landscape. Experts can be wrong—take for example the wrongheaded policies of fire suppression in national forests, and exotic species introductions to prevent erosion. In *The Geography of Childhood: Why Children Need Wild Places*, Gary Nabhan and Steven Trimble (1994) argue that children in particular crave and require direct access to little bits and pieces of nature around them every day. Large, pristine nature preserves even within short driving distance are insufficient.

Conservation of biological diversity across the entire landscape and from creek to ocean will not be accomplished by simply planting a few patches of shrubs and grasses on farms for "wildlife habitat." It will mean changing the face of agriculture itself.

Recovery of endangered animals such as the red cockaded woodpecker in the southeastern United States and the spotted owl and coho salmon in the Northwest have forced the U.S. Fish and Wildlife Service to confront the private lands issue in timber and ranching country. Experimental partnerships are forming among private landowners (some small and some very large, such as Weyerhauser), community planning groups, and government environmental agencies. But timber and ranching are extensive—rather than intensive—land uses involving fewer total landowners and less capital investment per unit area. Agriculture may be a harder nut to crack.

Who's in Charge of Conservation?

It is fair to ask, isn't somebody already taking care of conservation in farm country? The answer is of course yes and no. Land owners are a diverse lot, with many a concerned advocate for conservation of native biodiversity among them. But there are significant roadblocks to progress.

For the most part, rural people have not been willing or able to lead the way in defending environmental quality and biological diversity in their own communities. Farming is a rapacious business. The number of farms in the United States has decreased by 60 percent since 1964 (NASS 1999). In addition to financial stress, farmers face unpredictable weather, machinery breakdown, and disease outbreaks. There is not much room, economic or otherwise, to do anything on the land that doesn't pay for itself in cash relatively soon. In the Corn Belt and probably outside it too, fields are referred to as "ground," in the same way one might refer to standing board feet of timber, or tons of raw steel.

To make matters worse, farm organizations and commodity groups (e.g., the Corn Growers Association) tell their members that environmentalists are their worst enemy, bent on shutting them down via unsupportable regulation. Environmentalists are lumped with animal rights activists who oppose all livestock production. Scientific articles documenting problems such as global warming, the hypoxic zone in the Gulf of Mexico, and pollution caused by concentrated livestock are vilified in the pages of the *Farm Bureau Spokesman* purely on political grounds regardless of the quality of the research. Distrust and antagonism run rampant.

Rural communities are justifiably more worried about losing the farmers that support them than about losing the biological diversity of the countryside. One school board in the county with Iowa's highest average land values (translation: highest corn yields) asked a high school chemistry teacher to stop his students from monitoring the creek water running through farmland because they were finding too much contamination. Even when rural communities do summon the political will to fight for environmental quality, they have been barred from passing stricter environmental standards than state provisions, effectively limiting environmental decision making to the state level.

County Conservation Districts were set up across the country in the late 1930s in response to the dust bowl. Farmer-led district councils pass out federal money to plant trees along stream banks and build farm ponds. The winner of the 1999–2000 Grundy County (Iowa) Conservation Award farms 1,780 acres of corn and soybeans, including 70 acres of seed corn, using "no-till" farming methods. The soil is minimally disturbed during planting, and a large amount of crop residue is left on the soil surface over winter, preventing much of the soil erosion associated with more conventional practices—so far, so good. But no-till relies almost exclusively on chemical herbicides for weed control rather than field cultivation, and rates of nitrogen fertilization and leaching are the same as for conventional tillage. And only 27 acres—one and a half percent of the farm—isn't planted to row crops. This land consists of waterways seeded to domesticated grasses, and stream banks planted to trees and shrubs. Clearly, "conservation" in this context has nothing to do with nature.

At the state level, departments of natural resources or environmental protection tiptoe cautiously so as not to offend powerful commodity interests. Elected politicians owe their campaigns to the Farm Bureau and the National Pork Producers Council; state environmental rules are as lax as the federal rules will allow, and in practice more so. The U.S. Environmental Protection Agency has only recently begun to enforce the Clean Water Act's provisions for nonpoint source pollution from farms and urban runoff.

At the federal level too, the historic emphasis has been on soil conservation for the purpose of preserving the productive capacity of the land for continued farming. The United States Department of Agriculture's Soil Conservation Service (SCS) began by advocating fertilizer application to restore worn-out soils and compensate for the new practice of monocropping (Allaway 1957). The SCS provided technical and monetary assistance to farmers for introducing exotic species such as multiflora rose and kudzu for erosion control, draining wetlands, channelizing creeks, and building dams. They also promoted terraces on sloping ground, contour plowing methods, and conservation tillage (leaving the stubble from previous years' crops on the soil surface) to slow runoff. Aldo Leopold's bemused comment on the alternating speeding up and slowing down of water was that "the water must be confused by so much advice" (Leopold 1966, 126).

The conservation mission of the SCS began to expand in the latter half of the 1980s and in 1993 changed its name to the Natural Resources Conservation Service (NRCS). Only now are NRCS offices across the country beginning to train their engineering-oriented personnel to recognize native, relatively undisturbed habitats on farm land and make more sensitive recommendations about restoration, management, and conservation. There is a great opportunity for people with expertise in some group of wild animals or plants to become part of this agency and assist in its transformation.

It is tempting to dismiss national organizations such as Ducks Unlimited and Pheasants Forever because of their historic narrow focus on game birds. However, both have marshaled the considerable economic clout of hunters to buy and protect land, restore wetlands and upland habitat, and arrange partnerships between landowners and state and private conservation agencies. The U.S. Fish and Wildlife Service is also in a position to encourage conservation of biological diversity in farming landscapes. After all, this is the agency charged with listing endangered species and plotting their recovery. There are many uphill battles. The Upper Mississippi National Fish and Wildlife Refuge must compromise its conservation mission due to the powerful influence of the barge industry, which primarily moves grain and agricultural chemicals up and down a dammed and dredged river. In chapter 13, Carol Shennan describes the efforts of one national wildlife refuge to merge conservation efforts with economically sound potato production in California.

The Elephant in the Living Room

So far I have argued that the future of biodiversity in farm country depends on the future of agriculture. The largest store of knowledge of native biodiversity in farm country rests with field biologists at regional colleges and universities. Trained as mammologists, ornithologists, entomologists, botanists, and ecologists, this group has concentrated on studying and teaching *within* protected areas but for the most part has chosen not to get involved in the larger societal issues surrounding the industrialization of agriculture. I remember my undergraduate ecology class in south-central Iowa twenty years ago. We would disgorge from the college van into some remnant prairie along a railroad right-of-way, learn the names of the native plants, then get back in the van and drive away. We were trained to do biology "inside the box" of the prairie remnant and inside the box of classical field biology. Any reference to the farming landscape matrix was like mentioning the proverbial elephant in the living room—so big, so intractable and seemingly out of our control that perhaps our professors were afraid to broach the subject.

Things haven't changed much. A recent North American Prairie Conference featured presentations about plant communities in various prairie remnants, techniques for collecting and handling prairie seed, a discussion of genetic issues in prairie reconstruction, and field trips to local remnants and reconstruction sites. There is a great deal to learn about each of these topics, and much progress has been made. But the elephant is still in the living room, bearing down on each remnant or reconstructed prairie, stream, forest, and wetland. What is missing is a frank acknowledgement of the pressures that continue to threaten natural habitats and a willingness to confront them head-on.

I see several basic issues or misconceptions that have prevented restoration ecologists and conservation biologists from stepping up and becoming part of the conversation about land use in agricultural zones as Aldo Leopold once did in Coon Valley (see Tex Hawkins's account in chapter 4). First, the human notion of private property rights is at odds with the biological realities of wild animals and plants. Second, conservation biologists have uncritically accepted the idea that this land must be sacrificed to "feed the world." And finally, many of us are still unable to view agriculture as part of nature because of our roots in "pure" biology.

Private Property

Field biologists like to drive down lonely roads with one arm hanging out the window, spot interesting things, stop the car, and jump out to take a look. In the western United States, this is often possible without breaking any law or making anyone mad. A great deal of ranching and timber land is in some form of public ownership, and as long as you close the gate behind you, no one seems to mind. Later intensive collecting or experimentation might require formal permission, but the wide-ranging, exploratory phase of research is carefree and fun. In the Midwest, such access to land, even seemingly wild land, is restricted by the fact and culture of private property. Not only is most land private, but also more people *care* that it is private.

In farm country, it is impossible to explore and get to know a landscape without painstaking research of county plat maps and without letters and phone calls ahead of time requesting permission to trespass. As a result, research is often limited to university campuses or to state and county parks. It takes a special effort to befriend farmers and even more time to establish the trust and communication needed to do research on their land.

But the real problem is deeper. Many organisms (for example the bobolink, which migrates each year from wintering grounds in southern Argentina to nesting territory in the prairies of the United States and Canada) and most ecosystem processes (fire, water, energy, and nutrients) move across the ownership boundaries humans create. Conservation biologists are caught in the middle, understanding private property rights from a human perspective yet knowing that biologically speaking, whether we like it or not, the world is a commons.

Dean Kleckner, former president of the American Farm Bureau, correctly perceived the radical, essentially property-less nature of conservation in his criticism of the Endangered Species Act when the Associated Press quoted him in 1993 as saying: "Bats, bugs, chubs, suckers . . . they're claiming title to our land." Indeed, these animals have no idea what a title is and do not read no-trespassing signs, so any laws that protect the animals also challenge human property rights. Similarly, the Clean Water Act, which limits the destruction of wetlands and imposes maximum pollution levels for streams, has been threatening to landowners because it places restrictions on them and reduces their property values (Freyfogle 1995) In fact, the nature of property is not as simple and "real" as it may seem, and real estate law continues to struggle with instances such as the shifting channel of a river, in which nature will not hold still enough to belong to any one person or entity (Steinberg 1995). Conservation easements, which "unbundle" development rights from other kinds of property rights constitute one step in the process of freeing the popular imagination from the monolithic view of land ownership as an all-or-nothing proposition.

The contrast between human construct and ecological reality has created much unproductive tension. The Wildlands Project has developed breathtaking, comprehensive regional maps depicting huge wilderness preserves with broad buffers of restricted land use and corridors for animal movement. The conservation biologists who plan these extensive wildlands have avoided private property whenever possible, but they inevitably impinge on private property owners. Proponents view these plans as an "unflinching vision" of the actual requirements of the continent's largest carnivores (Soulé 1999/2000), but they could also be seen as an overt act of aggression against landowners.

Taking land for protected areas has a bittersweet past rooted in colonialism as well as nature worship. John Muir's beloved Yosemite Valley was used (and presumably loved) by Miwok people for hunting and gathering purposes for thousands of years. Similarly, Yellowstone National Park was used by Shoshone, Crow, Bannock, Blackfoot, and Sheepeater peoples. African national parks in Kenya, Tanzania, Zimbabwe, and South Africa dislocated rural people and disrupted their means of living. "Under colonial conservation laws, the collection of fuelwood became wood theft, the hunting of animals became poaching, and pasturing cattle became grazing trespass" (Neumann 1998, 34). Nature preservation, although "visionary," was a way for the colonial state to seize land and resources.

Given the politically charged struggles for use and preservation, it is no wonder that conservation biologists have, for the most part, avoided farming landscapes and stuck to public lands. Recent attempts to break the either-or stalemate with conservation easements, habitat agreements, and other incentives hold great promise for nature preservation on private property.

The Mandate to Feed the World

Conservation biologists may also avoid farming landscapes because it would appear craven and selfish to starve millions in order to preserve biodiversity (by keeping good farmland out of production and cutting back on fertilizers and pesticides). This was the main message of a publication entitled *How Much Land Can Ten Billion People Spare for Nature?* (Waggoner 1994) published by the Council for Agricultural Science and Technology. Monsanto (Horsch and Fraley 1998) and ecologist Michael Huston (1993) have suggested that already-cultivated areas be farmed even more intensively in order to feed the growing human population, estimated to top 10 billion before leveling off around the year 2050. It is hard to shirk this moral burden. In a land so wonderfully endowed with agricultural productivity and infrastructure, who would begrudge its sacrifice for the larger good of preventing world hunger?

But the argument is too simplistic. People go hungry for a variety of reasons unrelated to total agricultural output (Smil 2000). Agricultural output is capable of feeding all 10 billion people right now, if everyone in the world would eat a simple diet consisting mainly of grains, legumes, and dairy products. Losses during harvest, storage, transportation, cooking, and even at the table have a big effect on the total amount of calories and protein available to people. The conversion of farmland to housing developments and malls reduces potential food production more surely than the restriction of pesticides. Moreover, environmental destruction caused by agricultural intensification also reduces the potential food supply—for instance, when terrestrial nutrient pollution ruins shrimping in the estuary or fishing on the reef (Smil 2000, Matson et al. 1997, Vitousek et al. 1997). Agricultural productivity must contribute to an overall strategy to feed the world but should not bear the entire burden (Crofts 1987).

Conservation biologists and others who love and protect wilderness must not shrink from asking how human beings are to live. As good as author Edward Abbey was at articulating the need for wilderness, he never questioned where his beer came from. Many such pleas for "unmanaged wilderness" (e.g., Willers 1999) are devoid of any reference to human food and how it will be grown. The food system (and fiber, and wood systems) discourages this line of questioning. Food conveniently comes from a grocery store or a restaurant, and it is cheap by middle class standards. Its provenance is distant and mysterious, the people who grow it are faceless, and we as consumers are kept blissfully ignorant of the ecological consequences of our very appetites.

Pure Versus Applied Biology

Ultimately, conservation biologists' reluctance to look at farming landscapes comes from the tradition of "pure" natural history studies in the United States (but not the United Kingdom; see chapter 9). Agriculture is the consequence of will and purposeful action, while nature is "wild" or "self-willed." In the minds of many a conservation biologist and backcountry hiker, a farm is no more natural than a parking lot. But whenever one looks past the square fields and giant tractors of the present, one sees agriculture as an extension of fundamental biological processes common to all creatures. For example, the complex ecological interactions between crops, weeds, pests, and soil in the corn-bean-squash polyculture of Central American peasants are clearly "natural" though managed by people. In modern fields, wild deer fatten on part of the crop and are slaughtered and processed alongside the tame cattle and hogs in local meat lockers. The genes of wheat mingle with those of wild goatgrass on field edges. Free-living worms and soil microbes feed on crop residues and animal manures, turning them into valuable topsoil. As any farmer will tell you, a lot of what happens on farms is beyond human control.

The Way Forward

The task before us then is to integrate, in our minds and in our farming landscapes, the wild and the willed. We need to help devise ways of reconciling the needs of wild animals and plants with a working landscape full of human cultigens, boundaries, and institutions. We need to interact directly with farmers, searching for common ground as Aldo Leopold did before us. This will mean being willing to learn how farmers' decisions are constrained by markets, agricultural policy, history, labor, and capital as well as by natural resources. Ultimately, it will mean coming to terms with the consolidation of land and markets into fewer, more powerful hands. "The farmer," able to make decisions independently for the good of family and community, is being replaced by people who work for land management companies, contractors growing livestock for large corporations, custom manure haulers, and hourly wage tractor jockeys. Our conversations with disenfranchised rural residents about the future of the land and its wild inhabitants will have substantially less promise of success than our efforts do today.

References

Allan, J. D., and A. S. Flecker. 1993. "Biodiversity Conservation in Running Waters: Identifying the Major Factors That Threaten Destruction of Riverine Species and Ecosystems." *BioScience* 43:32–43.

Allaway, W. H. 1957. "Cropping Systems and Soil." Pp. 386–395 in *Soil: The 1957 Yearbook of Agriculture,* edited by A. Stefferud. U.S. Department of Agriculture, Washington, D.C.

Avery, D. T. 1995. *Saving the Planet with Pesticides and Plastic: The Environmental Triumph of High-Yield Farming*. Hudson Institute, Indianapolis.

Berendse, F., R. Aerts, and R. Bobbink. 1993. "Atmospheric Nitrogen Deposition and Its Impact on Terrestrial Ecosystems." Pp. 104–121 in *Landscape Ecology of a Stressed Environment*, edited by C. C. Vos and P. Opdam. Chapman and Hall, London.

Crofts, T. 1987. *The Return of the Wild: The British Countryside and the World-Wide Rural Crisis*. Friendly Press, Oxon, U.K.

Edgerton-Warburton, L. M., and E. B. Allen. 2000. "Shifts in Arbuscular Mycorrhizal Communities Along an Anthropogenic Nitrogen Deposition Gradient." *Ecological Applications* 10:484–496.

Freyfogle, E. T. 1995. "The Owning and Taking of Sensitive Lands." *UCLA Law Review* 43:77–138.

Horsch, R. B., and R. T. Fraley. 1998. "Biotechnology Can Help Reduce the Loss of Biodiversity." Pp. 49–65 in *Protection of Global Biodiversity: Converging Strategies*, edited by L. D. Guruswamy and J. A. McNeely. Duke University Press, Durham, N.C.

Huston, M. 1993. "Biological Diversity, Soils, and Economics." *Science* 265:1676.

Leopold, A. 1966. *A Sand County Almanac with Essays on Conservation from Round River*. Ballantine Books, New York.

Matson, P. A., W. J. Parton, A. G. Power, and M. J. Swift. 1997. "Agricultural Intensification and Ecosystem Properties." *Science* 277:504–508.

Nabhan, G. P., and S. Trimble. 1994. *The Geography of Childhood: Why Children Need Wild Places*. Beacon Press, Boston.

National Agricultural Statistics Service (NASS). 1999. *1997 Census of Agriculture-United States Summary and State Data*. Geographic Area Series Part 51, Vol. 1. U.S. Department of Agriculture, Washington, D.C.

Natural Resources Conservation Service (NRCS) 1996. *America's Private Land: A Geography of Hope*. U.S. Department of Agriculture, Natural Resources Conservation Service, Washington, D.C.

Neumann, R. P. 1998. *Imposing Wilderness: Struggles over Livelihood and Nature Preservation in Africa*. University of California Press, Berkeley.

Schneiders, R. K. 1999. *Unruly River: Two Centuries of Change Along the Missouri*. University Press of Kansas, Lawrence.

Smil, V. 2000. *Feeding the World: A Challenge for the Twenty-First Century*. Massachusetts Institute of Technology Press, Cambridge.

Soulé, M. E. 1999/2000. "An Unflinching Vision: Networks of People for Networks of Wildlands." *WildEarth* 9:38–46.

Soulé, M. E., E. T. Bolger, A. C. Alberts, J. Wright, M. Sorice, and S. Hill. 1988. "Reconstructed Dynamics of Rapid Extinctions of Chaparral-Requiring Birds in Urban Habitat Islands." *Conservation Biology* 2:75–92.

Steinberg, T. 1995. *Slide Mountain, or the Folly of Owning Nature*. University of California Press, Berkeley.

Tilman, D., and D. Wedin. 1991. "Plant Traits and Resource Reduction for Five Grasses Growing on a Nitrogen Gradient." *Ecology* 72:682–700.

Vitousek, P. M., J. D. Aber, R. W. Howarth, G. E. Likens, P. A. Matson, D. W. Schindler, W. H. Schlesinger, and D. G. Tilman. 1997. "Human Alteration of the Global Nitrogen Cycle: Sources and Consequences." *Ecological Applications* 7:737–750.

Waggoner, P. E. 1994. *How Much Land Can Ten Billion People Spare for Nature?* Council for Agricultural Science and Technology Task Force Report No. 21. Ames, Iowa.

Willers, B., editor. 1999. *Unmanaged Landscapes: Voices for Untamed Nature*. Island Press, Washington, D.C.

Part II

Restoring Nature on Farms

The five chapters in Part II feature farmers whose practices are restoring nature on their farms. They are not all winning Conservationist of the Year awards for following soil conservation plans designed by the Natural Resources Conservation Service or being given recognition for creating special habitat for game birds. Their management decisions are not motivated just by cost-share programs available through various agencies of the U.S. Department of Agriculture, although such assistance is often appreciated. Such programs jump-started land recovery in the 1930s when the Soil Conservation Service was created, and they show promise of jump-starting wetland restoration and stream-protection efforts today, but long-term stewardship requires more than financial incentives.

Conservationists all over the world know Aldo Leopold as an articulate spokesperson for wilderness. They are less familiar with his equally strong belief that as well as having fields and pastures, working farms should have forests, marshes, and prairie, because we need wildlife habitat on private land. Leopold spent most of his life in the agricultural Midwest, and he did not accept agricultural landscapes as ecological sacrifice areas for food production.

Aldo Leopold's influence is obvious in this book, particularly in this section. Chapter 4 was written by Tex Hawkins, whose father was a student of Aldo Leopold, and chapter 5 by Buddy Huffaker, currently the executive director of the Aldo Leopold Foundation. Chapters 6, 7, and 8, by Brian DeVore, introduce readers to farmers that live Leopold's land ethic.

All five chapters describe farmers who are managing their land in ways that create harmony between farming and the natural world. The authors explore from several angles the motivations and accomplishments of individual farmers in conserving soil, water, and wildlife. Though economics have sometimes forced them to experiment with low-cost inputs and farming practices, they have been driven mainly by personal or family goals and values.

Tex Hawkins is a wildlife biologist employed by the U.S. Fish and Wildlife Service. Arthur Hawkins, his father, was a student of Aldo Leopold in the 1930s, when the Coon Valley Watershed Project was established to reduce soil erosion on cropland and sediment loads in the Mississippi River. In chapter 4, "Return to Coon Valley," Tex describes a trip to Coon Valley, Wisconsin, where Leopold had sent his father sixty-three years earlier to examine the impact of land restoration on wildlife. He visits with two families whose farms were among the 418 in the Valley that made agreements to partner with the government to restore their land, and he learns that they still follow the original conservation plans even though the partnership has long since ended. And, although today's agricultural policy and farm economy do not reward conservation farmers for their current efforts, the agroecological restoration that resulted from those early conservation plans is still quite evident. Tex speculates on the future of the watershed and the Mississippi River ecosystem in the face of ever-increasing crop specialization and the trend toward consolidating land ownership.

In chapter 5, "Reading the Land Together," Buddy Huffaker, executive director of the Aldo Leopold Foundation in Baraboo, Wisconsin, explains how the Foundation works in the best tradition of its namesake to promote harmony between people and the land. Finding himself in the center of a conflict over the purchase of farmland for the purpose of establishing a national wildlife refuge in the vicinity of the Aldo Leopold Shack, Buddy was thrust into a leadership role to help resolve differences between angry farmers and government employees. He describes how a local committee of farmers and conservationists developed plans for farmers to enhance wildlife habitat on privately owned farms rather than allow the government to purchase it. Buddy writes about other efforts of the Foundation to blend private and public projects to restore natural habitat in the region, in particular oak savannah restoration along the bluffs of the Mississippi River in Wisconsin. In the Foundation's experience, the involvement of local citizens in planning and decision making for habitat improvement is just as important as government incentives.

Chapters 6, 7, and 8 are by Brian DeVore, editor of the *Land Stewardship Letter*, the official publication of the Land Stewardship Project. He introduces a number of exemplary farmers he has interviewed whose farming practices have evolved to be more in harmony with the natural world. In chapter 6, "When Farmers Shut Off the Machinery," DeVore writes that the habit of paying attention, shutting off the machinery, and walking the land to look and listen guides some farmers in making decisions that work for them, even though they may be contrary to conventional wisdom. In chapter 7, "Stewards of the Wild," we meet two Iowa farmers and a South Dakota farmer who pay attention to what happens to water that falls on their land. The first Iowan monitors water quality in a pasture creek that flows through a fen, and he treasures woods and prairies on his land. The other, an avid fisherman, keeps his steep land in permanent pasture and a mix of rotations and cover crops to prevent soil and nutrients from finding their way to the Mississippi River nearby. The farmer in South Dakota establishes temporary check dams in the spring to provide nesting habitat for waterfowl and finds that wetland vegetation filters out excess nutrients from water flowing through it into a lake below. In examining why farmers manage their land differently from neighbors on conventional farms, DeVore finds in chapter 8, "Why Do They Do It?" that in addition to personal land ethics, such farmers make the necessary ecological connections. They see the need for systemic changes, in contrast to conventional farmers who tend to look for single solutions to solve what they see as single problems.

The farmers described in this chapter enjoy the natural habitats on their farms, and they see them as part of the farm, just as the farm is part of a larger ecosystem. Such understanding reinforces a land ethic that tells them it is their responsibility to conserve the soil, the water, and the wild, whether or not the government pays them to do so. Leopold would be pleased.

Chapter 4

Return to Coon Valley

Arthur S. (Tex) Hawkins

In a 1935 article for *American Forests*, Aldo Leopold described Coon Valley, Wisconsin, as "one of a thousand farm communities which, through the abuse of its originally rich soil, has not only filled the national dinner pail, but has created the Mississippi flood problem, the navigation problem, the overproduction problem, and the problem of its own future continuity" (Leopold 1935).

When Leopold wrote this article, "An Adventure in Cooperative Conservation," his sons Luna and Starker were working in Coon Valley with the newly established Soil Erosion Service on the nation's first watershed project. Farmers and natural resource technicians were working as partners to develop comprehensive conservation plans and restoration methods. Leopold felt that this project would be the prototype for restoring ecological health and economic prosperity to the dust bowl–ravaged landscape. He was particularly interested in documenting how wildlife species would respond to profound changes in agricultural land use. He sent a graduate student who had been studying quail population irruptions to Coon Valley for a closer look. Ten years later, the student would become my father, and sixty-three years after he visited Coon Valley, I found myself driving on a better road to the same destination with a parallel mission. I too would visit with local farmers and other natural resource professionals to assess wildlife's response to land use changes and also look at the overall condition of the land after more than a half-century of cooperative conservation. And I would see what I could learn about restoring nature on farms.

Nation's First Watershed Project

It was a drizzly October morning when I turned southeast from LaCrosse, Wisconsin, on Highway 61. I left the Mississippi River and ascended a valley known as Mormon Coulee for several miles before crossing a cloud-shrouded ridge and descending into a scene that seemed to belong on an old-fashioned wall calendar. Against the green wooded bluffs, I noticed that a few maples and sumac clumps were beginning to turn yellow-orange. Strips of lush alfalfa contrasted with dry standing corn. Holstein cattle grazed near a freshly painted red barn. A flock of pigeons circled the steeple of a white church.

The sleepy but prosperous village of Coon Valley stretched out along the highway—shops and service stations facing the traffic with streets and homes neatly tucked in behind. A town park with a handicapped fishing access had recently been completed by partners (recognized with a monument) on the bank of Coon Creek. Across the naturally restored stream was a level field where the Civilian Conservation Corps (CCC) barracks had stood back in the 1930s. I crossed the new highway bridge over Coon Creek and pulled off at a historic wayside to read a sign with the title, "Nation's First Watershed Project."

The sign explained that this 90,000-acre drainage basin was selected in October 1933 by the Soil Erosion Service (which later became the Soil Conservation Service and is now called the Natural Resources Conservation Service [NRCS]) to be the first large-scale demonstration of soil, water, forestry, and wildlife conservation measures to help restore the vitality and productivity of farming. Using pooled experience and knowledge from the University of Wisconsin and agency technicians, local farmers were assisted in establishing "a pattern of land use now prevalent throughout the Midwest."

Obviously, I thought, there have been changes in the overall pattern since *that* line was written, even though it's just as obvious that Coon Valley still attests to the enduring "wisdom, courage and foresight of farm families" who succeeded in repairing damages wrought by drought and land abuse that culminated in the Great Depression.

But how successful was the Coon Valley Project? How widespread and lasting were the benefits? Do the improvements we see in today's landscape, compared to historic photos of gullies and floods deep enough to hide a house, justify the time or expense that has been invested by so many over the years? Does the work continue today, or are positive trends now being gradually reversed?

These questions were on my mind when I transferred my gear into the pickup of Vernon County's NRCS district conservationist Jim Radke. He explained to me, as we left town on a winding gravel road, that the watershed project's original plans assigned land use based on slope, which made sense to me, considering the steep terrain. Starting at the top of the bluffs, 0 to 20 percent slopes were cropped, 20 to 30 percent slopes were pastured, and slopes over 30 percent were considered best suited to woodland. At the base of the bluffs, slopes from 20 to 30 percent were pastured, and under 20 percent were cropped right down to the creek. Plans called for contour farming and strip crops in rotation, reforestation with selective tree harvest, a variety of practices for gully and stream-bank repair, and winter cover for wildlife. CCC assistance was provided for fencing, terracing, water controls, pasture seeding, and fertilization to restore basic productivity. The government provided jobs, wages, and skill training for workers at the camp, who otherwise might have been unemployed and unable to provide for their families. Out of 800 farmers in the watershed, 418 signed on between fall 1933 and spring 1935 (Helms 1985). Farmers signed comprehensive five-year cooperative agreements that provided them with detailed plans for restoration (Helms 1985).

Agricultural economist Melville H. Cohee assisted local farmers with bookkeeping and evaluating financial progress, while government technicians and professionals pooled their expertise to develop field maps, plans, designs, budgets, and schedules. Cohee recalled that Aldo Leopold had

> spent many days and nights on the Coon Valley project in 1934 and 1935. In the beginning he practically engineered most of the wildlife management work for the project, showing how to make inventories of habitats and species of wild animals. . . . This most probably was Leopold's first opportunity to help put into actual operation many beneficial practices. . . . It was Leopold who caused us young SES personnel to go beyond physical and economic considerations and into the ethical and ecological deeper reflections on what our program could offer farmers, the community and others. (Cohee 1987)

When plans were approved, crews went to work putting informed decisions of each participating landowner into practice. Some of the improvements and benefits, such as small check dams to control runoff, were immediately obvious. Others, such as habitat plantings and brush piles to provide winter wildlife cover, probably looked untended and may have taken longer for landowners to adopt. But the overall trend was toward better integration of production and protection. Farmers who laid out contour strips and terraces and agreed to test the system for five years were supplied with seed, lime, and fencing (Leopold 1935).

As a result of the combined efforts of farmers and natural resource specialists in the Coon Valley Watershed, erosion rates from participating farms were gradually reduced to about one-quarter of the dust-bowl levels, as stream hydrology improved (Trimble and Lund 1982).

The Semke Farm

Jim Radke pulled off the ridge road onto a driveway lined with miniature orange pumpkins. We parked between the Kevin Semke family house and barn. As a friendly dog trotted out to greet us, Jim explained that the farm's original owner, Lester Mundstock, had been among the first cooperators on the Coon Valley Watershed project. Lester Mundstock died in 1990, but his wife Elnora still displays their Outstanding Conservation Farm plaque on the kitchen wall. The family conservation tradition continued when Elnora's daughter Doloris and her husband Gene Schmeckpeper took over the farm. And now *their* daughter—Elnora's granddaughter—Ellen and her husband Kevin Semke own and operate the farm, representing the fourth generation to maintain essentially the same contour strip rotations that were laid out in 1933.

According to Kevin, the three Semke children have learned the conservation strategies used by their predecessors essentially through osmosis—by seeing and doing things "the right way." The kids "just know" how to prevent soil loss by modifying tillage methods when this is necessary to protect unstable slopes. Kevin showed us the location of a shack in a restored tract of woodland on the map where he retreats during deer hunting season, and he proudly shared articles describing Future Farmers of America awards and accomplishments of all three children. Clearly, twin traditions of appreciation for the outdoors and recognition of achievement have helped sustain this farm, I thought.

I asked Ellen if her grandparents were at all reluctant to enter into an agreement with the government. She replied that people had warned them not to sign because government might take the farm, but her grandfather had said, "We have nothing more to lose."

So here they are, four generations into the process, and it seems like Lester Mundstock made the right decision. The family sustains economic benefits by constantly exercising esthetic and ethical judgement and by constantly reinvesting in the maintenance of conservation practices, which would otherwise tend to deteriorate over time.

Jim Radke observed that the true depth of a stewardship ethic is demonstrated not so much by what people say, but by the long-range commitments that they make. "This family routinely makes voluntary improvements on rented property as well as their own, and even the kids make an effort to share ideas and information with neighbors who are just beginning to implement more sustainable practices."

Aldo Leopold anticipated that project benefits would be about as lasting as the commitments of local individuals to the shared goals of the partnership. In other words, conservation challenges would best be met in Coon Valley, and elsewhere, through a revolutionary "land ethic" (Leopold 1949).

Back on the road, I worried aloud that this kind of family-scale diversified dairy farming may not be as sustainable as we might hope, in view of trends in farm size, markets, and prices. Radke agreed, noting that Vernon County had lost about 250 dairy farms, or about one-fifth of its total, over the previous eight years.

This trend has continued. As dairy farms shrink in numbers and grow in size, there has been a corresponding increase in feed grain production, often on land better suited to less erosive crops, hay, or pasture. Over a two-year period, Radke reported a 20 percent increase in corn and a ten-fold increase in soybean acreage. The conversion was not then as complete as it had been in other areas, but it was underway. I reasoned that this continuing concentration of land ownership and increasingly intensive row cropping, more dependent on chemical fertilizer, pesticide, and fuel-guzzling machines than were the original diversified family farms, could reverse soil-loss trends if drought returned.

Sediment Control and the Mississippi River

During the past half-century, rates of sediment eroding off cropland in the Coon Creek watershed have declined significantly. Significant improvements have been observed throughout the remainder of the Mississippi River basin as Soil and Water Conservation District and NRCS employees have helped conservation-minded farmers implement practices (Trimble 1999).

Thanks to the pioneering research of scientists like Stafford Happ, Stanley Trimble, James Knox, and others, a clearer picture has emerged of the catastrophic impacts to the river ecosystem that led up to the dust bowl, and the amazing healing process that followed. By surveying cross sections of tributary valleys and studying sediment cores taken over a period of decades, the investigators helped demonstrate how the intensification of agriculture since settlement produced enormous amounts of floodplain erosion and sedimentation (Waters 1995).

Trimble's latest analysis shows that the amount of sediment discharged by Coon Creek into the Mississippi River remained fairly constant over the twentieth century, even though erosion rates in the watersheds have gone through dramatic changes. The reason for the close to constant sediment yield, as measured at the watershed's point of discharge into the Mississippi, is the deposition within the watershed. During the period of maximum soil erosion rates, from approximately 1920 to 1940, the flood plains of the lower main valley received about 15 centimeters of vertical accretion of sediment a year. Rather than discharging directly from the mouth of Coon Creek, suspended sediment settled in the alluvial flood plain. Following improvements in farmland management, the rate of sediment accretion in the lower valley fell to .53 centimeters a year from 1975–1993. While overall rates of sediment storage decreased, the sites of sediment storage also shifted to flood plains higher up in the watershed. Some of these flood plains were newly developed, the result of lateral stream bank erosion that widened channels and created new, wider flood plains that became vegetated and served as sediment traps. Trimble's research demonstrates that deposition patterns of the Driftless Area are dynamic. Therefore caution should be used in relating sediment yield to rates of soil erosion (Trimble 1999).

At first, conversion of prairies to wheat fields, and woodlands to pastures, meant greatly accelerated runoff and erosion from the steep terrain surrounding tributary streams. During the drought, severe floods became so frequent that farms and even entire communities around milling sites had to be abandoned. Within a few decades, 10 to 30 feet of topsoil and heavier sediments covered the broad flood plains of major streams, burying roads, bridges, and fields. In some places, barns were buried to their roof lines, leaving only the tops of their silos showing. Clearly, land use changes were producing costs that outweighed benefits. The nation's precious topsoil, dislodged by the drought, was being gradually flushed down streams to smother fish spawning areas, clog the navigation channels of the Mississippi River, and eventually discharge future generations' wealth into the Gulf of Mexico (Thieling 1998).

Something had to be done, and the first place to look for solutions back in those days was at higher technology and heavier engineering. Lateral deflectors known as wing dams had been used since the late 1800s to concentrate the Mississippi's flow and thus deepen the navigation channel. By the 1920s, it was clear to decision makers that the Mississippi, like most other great rivers, needed dams. Engineers developed plans for a series of locks and dams along the of the Upper Mississippi River to create navigation pools sufficiently deep to float loaded barges, even during drought. Vast portions of the flood plain would be diked to further increase the amount of land available for crop production (Hoops 1988).

When river enthusiasts heard about the plans, they were shocked. Will Dilg, an advertising executive from Chicago, joined with other outdoor enthusiasts who had long enjoyed hunting and fishing in the Mississippi River backwaters to form the Izaak Walton League. Through a sophisticated membership and lobbying campaign, a powerful conservation organization was created almost overnight. Dilg actually paid recruiters a bounty for new members, and "Ikes" chapters were quickly created in nearly every state. He set up his operations headquarters in a Washington, D.C. hotel, and after marathon sessions with members of Congress, got legislation passed in 1924 to establish the Upper Mississippi River National Wildlife and Fish Refuge, which today protects about 200,000 acres of backwater habitat along 260 miles of river (Moore 1998).

Much of that habitat was flooded or otherwise transformed during the 1930s as the locks and dams were completed. But there were positive outcomes as well. Near the middle of the navigation pools, large, productive marshes were created just when migratory waterfowl were being displaced from drained and drought-stricken prairies of the four neighboring states. For a quarter-century, fish and wildlife prospered and outdoor recreation expanded. But the process now known as reservoir aging eventually caught up. At first, few people noticed the accumulation of sediment beneath the water's surface. Then, mud flats and deltas began to expand, even as floods scoured some channels deeper. Contamination grew apace with urban sprawl and industrialization. When plant growth was stunted by turbidity from wind and wave action and the feeding and spawning activity of rough fish like carp, both scientists and recreationists became alarmed (U.S. Fish and Wildlife Service 1987).

Collaborative Conservation

During the 1970s, a state/federal interagency effort known as the Great River Environmental Action Team (GREAT) was formed to study mounting threats to habitat, recreation, and navigation and to make recommendations for future management. A comprehensive basin management plan was completed in 1982. It called for designation of dredge disposal sites to maximize recreation opportunity and to minimize backwater impacts. Using radioactive trace elements, soil scientists were able to measure the rate of sediment accumulation in several locations. They predicted a life expectancy of fifty years or less for many of the Mississippi's most productive and popular backwater areas. They also estimated that it could cost the federal government over a billion dollars over a twenty-year period to tackle erosion and sedimentation problems at the source (Upper Mississippi River Basin Association 1984).

Individually, no government agency could address problems of this magnitude, since all were effectively competing for little pieces of the same limited pie. Consequently, the agencies delayed tackling the number one threat to the Mississippi River habitats, which was identified in the GREAT studies as sediment accumulation. This did not mean that individual agencies sat on their hands. On the contrary, most tried to work within their mandates to attack different aspects of the sedimentation problem with existing program and personnel.

For example, the 1985 Farm Bill and subsequent updates have created stronger conservation incentives through many U.S. Department of Agriculture programs. On the floodplain of the Mississippi, between the railroad tracks and highways that traditionally define "Upper Miss" refuge boundaries, the U.S. Fish and Wildlife Service coordinates management activities with the Corps of Engineers and the adjoining states. This agency also participates in pilot watershed management initiatives through its newly expanded Partners for Fish and Wildlife Program, which has supported my participation in local projects since 1987.

Collectively, agencies and private organizations are moving toward an ecosystem-based approach that is adaptive, holistic, and long range. This approach requires that participants view management challenges broadly and understand the interrelatedness of issues. To put it another way, the scale of analysis has to reach beyond political boundaries to encompass ecological entities—such as watersheds—while the scale of implementation has to reasonably respond to individual or community priorities and freedoms. This would seem paradoxical, if it weren't for the amazing foresight and insight on today's political predicament that Aldo Leopold provided more than fifty years ago.

> Lack of mutual cooperation among conservation groups is reflected in laws and appropriations. Whoever gets there first writes the legislative ticket to his own particular destination. We have somehow forgotten that all this unorganized avalanche of laws and dollars must be put in order before it can permanently benefit the land, and that this onerous job, which is evidently too difficult for legislators and propagandists, is being wished upon the farmer and upon the administrator of public properties. (Leopold 1935)

Conservation on Private Land

Having neatly captured the essence of the bureaucratic snarl and gridlock that continues to impede some conservation efforts today, Leopold proceeded to define what makes the example of Coon Valley as applicable today as it was during the dust bowl:

> The farmer is still trying to make out what the many-voiced public wants him to do. The administrator, who is seldom trained in more than one of the dozen special fields of skill comprising conservation, is growing gray trying to shoulder his new and incredibly varied burdens. The stage, in short, is set for somebody to show that each of the various public interests in land is better off when all cooperate than when all compete with each other. This principle of integration of land uses has been already carried out to some extent on public properties like the national forests. But only a fraction of the land, and the poorest fraction at that, is or can ever become public property. The crux of the land problem is to show that integrated use is possible on private farms, and that such integration is mutually advantageous to both the owner and the public. (Leopold 1935)

The same year that Leopold published these words, he bought a worn-out and abandoned farm near Baraboo, along the Wisconsin River, and set about involving his family in the restoration of the land and the chicken coop that became known as "the Shack." Since the impoverished soils on their farm had never been very suitable for pasturing cattle or growing feed corn, the Leopolds restored their farm to native prairie, woodland, and marsh.

Today, the property literally sings praises to the Leopold family every day of the year. Breezes whisper through pine needles and rustle through oak leaves in the winter. Frogs and toads sing in the spring. Owls hoot on summer nights. Jays call and squirrels chatter while they plant the seeds for future pines and oaks each fall. The voices of nature tell us what can happen when Leopold's "land ethic" is put into practice on any marginal farm.

Aldo Leopold also taught ecological restoration principles to his graduate students in wildlife management at the University of Wisconsin. And during the 1930s, my father assisted "The Professor" while working on his master's thesis and writing "A Wildlife History of Faville Grove," documenting land management changes on a cluster of conservation farms just east of Madison (Hawkins 1940).

These farms were homesteaded by the Faville family in 1845. Stoughton Faville's daughter Ellen married Frank Tillotson, who eventually took over the farm, and their eldest daughter Betty married Art Hawkins, my father, in the summer of 1941. The ceremony was held on the family's unbroken prairie reserve, which still borders the Crawfish River north of Lake Mills.

About ten years later, my parents bought their own worn-out dairy farm beside marshy Lake Amelia, about fifteen minutes north of St. Paul. And for the past fifty years they have applied Leopold's lessons on land stewardship to restore ecological integrity to that portion of suburban Minnesota. Ospreys and loons now nest there each summer.

As far as I know, neither the Leopold nor the Hawkins farms ever turned profits producing commodities. But they did produce other forms of wealth and security. Countless natural blessings were shared among family members, while environmental and educational benefits were shared with surrounding communities. These benefits were locked in when my parents put their land under easement with the Minnesota Land Trust. The experiences that I had, growing up on a restored farm under the influence of Leopold and my parents, I suppose are what eventually brought about my return to Coon Valley.

The Haugen Farm

"Well, here we are finally at the Haugen Farm." Jim Radke interrupted my reflections as he parked the pickup in front of a small frame farmhouse sheltered by a row of stately Norway spruce trees. Smoke from a wood stove was rising from the chimney.

Ernest and Joseph Haugen greeted us in strong Norwegian accents, and we sat down at the kitchen table as they carefully unfolded the original copy of their 1934 conservation plan, which their father signed and they continue to implement, as written. The plan provided detailed instructions, with diagrams of pasture, forestry, and wildlife practices, such as shelter belts. Colored contour strips on five separate map pages clearly showed corn, small grain, and alfalfa rotations for all crop fields. Ernest pointed to a weathered chart specifying crop rotations on every narrow strip from 1949 through 1998. Then he showed us a list of seed supplied by the government under the plan: 543 pounds of alfalfa, 143 of timothy, 42 of red top, 42 of alsike, 138 of red clover, and 60 of Sudan grass.

The plan also provided for 1,200 rods of fencing along the contour, construction of a twenty-four rod terrace with a bulldozer, and application of 100 tons of lime and 8,500 pounds of fertilizer. "Back then," he said, "the farm was completely ruined and all full of gullies. We had to pile up grain bundles to get across the ravines." When some of the pasture plantings didn't take, additional seed was provided. By 1936, the combination of improved weather, plantings, and fertilization paid off. Their 34 acres of alfalfa "filled the barn."

The Watershed Project, Joseph said, meant a lot of extra work, especially spreading lime and fertilizer using a hand-crank cyclone seeder in the wind, with "sore eyes." But later, their father was asked how he liked the new terrace, and he replied, "Ask the folks down below who used to get flooded."

The Haugen brothers told about going down to the village of Coon Creek and seeing the flood waters crest, sometimes twice in the same day. "We used to go over to the Erickson Schoolhouse," Joseph recalled, "to see presentations put on by the CCC." In one play, he remembered with a grin, poor farming and gully erosion was described as "the quickest way to get to China."

The first major flood recorded by the village of Coon Creek was in 1874, just ten years after the Haugen farmhouse was constructed. At the peak, major floods were occurring almost every year, but since the watershed project, floods of that intensity have been reduced to roughly twenty-year intervals (Trimble 1982).

However, delivery of sediment to the stream and the backwaters of Pool 8 on the Mississippi, where Coon Creek finally deposits its load, continues to be a problem. According to Jim Radke, there are still perhaps 50,000 active gullies in Vernon County hidden by shrubs and trees in low spots along the edges of bluff top fields where runoff is concentrated. Stream-bank and streambed erosion during floods also continue to be major sources of sediment, as floodplain soil deposits are "mined" by water and carried downstream (Trimble 1999).

Radke explained that many erosion-prone fields had been enrolled in the Conservation Reserve Program under the assumption that re-vegetating slopes would be sufficient. But when farmers stopped cultivating to allow permanent vegetation to get established, dense shrubs or trees invaded and shaded out the grasses, which were needed to form sod and hold topsoil. Then, erosion began to scar steep slopes again, as gullies crept back up into the fields.

Ideally, he said, additional funding should be provided to repair gully erosion and maintain the grassy conditions that help prevent it. Practices could include annually clipped grass filter strips along field borders, as well as diversions, waterways, and grade stabilization structures to better manage flow from field runoff.

At present, limited federal and state funds are targeted into certain geographic areas for "best management practices" to address specific problems. Family farmers like the Haugens, who live in less degraded areas because they carefully follow their conservation plans, avoid problems by habit and therefore may not qualify for targeted cost-sharing assistance.

The very self-sufficiency of small, diversified farms puts them at a competitive disadvantage, since the good stewards voluntarily manage their land in ways that prevent problems from developing in the first place. Efforts that benefit society as a whole may go unrewarded, while recurrent damages resulting from unwise policies and poor stewardship qualify for financial assistance (Levins 2000).

For example, under the U.S. Department of Agriculture's Environmental Quality Incentive Program (EQIP), only 20 percent of Vernon County initially qualified as a priority area, and this designation didn't include the Coon Valley Watershed at the time of my visit. Consequently, funding for conservation practices bundled under EQIP was not available to the Haugens then, or to many of the other farmers who had followed their conservation plans religiously for generations.

"I didn't even know the Haugen brothers existed until a few years ago," Radke said. "They have been out here for all these years, following their conservation plan independently." As we walked out of the sparsely furnished farmhouse, I wondered why such a disproportionate share of farm conservation payments seemed to be going to the largest and wealthiest farms to correct chronic problems of land abuse instead of to the smaller family farms that clearly need assistance to sustain profitability and good stewardship of land and water.

On the way back to the truck, I heard soft songs of bluebirds—about forty were lining up on the fence wires and the power lines overhead. I noticed a rusty ham tin nailed to a fence post with a hole neatly cut out of the narrow part near the top. Ernest was smiling. "They like it here," he said. "And the bobolinks and meadow larks do too. I think it must be because we don't mow until later in the summer than most folks, and that gives some of the young birds a chance to get out of the nest." We talked about the continental decline in grassland nesting species, as well as other migratory songbirds, that has been documented from breeding bird surveys over the past twenty years. I told them how encouraging it was to see that so many of these species still reside on the Haugen farm, thanks to their efforts.

Coon Valley Today

As Jim Radke and I drove back down the winding gravel road into Coon Valley, I mentioned how impressed I was by the overall health of the landscape. Considering that experts judged topsoil loss to be over 75 percent throughout most of the region when the watershed project was initiated, recovery has been amazing.

"When we looked at historic air photos of stream reaches taken between 1939 and 1967," Radke replied, "we could see how channels meandering across the flood plains were moving dramatically from year to year. In contrast, photos taken from 1967 up through the present show less channel movement and bank erosion seems to be less severe, except at problem sites, where we're beginning to target some of our stabilization and restoration efforts.

"Of course, stream improvement is also a result of better cultivation and pasture management, which increase infiltration, while decreasing runoff and erosion," he added. "We have many farmers to thank for fencing to exclude cattle from eroding woodlands and stream banks, and for cultivating their fields on the contour with strip crops and conservation tillage.

"In terms of soil, water, and wildlife conservation here in Coon Valley," Radke said, "we're way ahead of where we would have been without the project. Three-quarters of the farms in this area were planned and treated under the original watershed partnership."

Since the 1985 Farm Bill, field checks have been used to verify compliance with measures required of anyone who receives federal assistance. Although the Conservation Reserve Program (CRP), contour strips and conservation tillage have greatly reduced sheet and rill erosion, and most of the area's woodlands are no longer grazed, Radke is quick to say "we've still got a long way to go . . . even on cropland that's contour stripped, we can document that the runoff rates are three times what they would be if permanently vegetated."

Just then the road straightened and we were back in Coon Valley. Jim Radke finished our tour by stressing that around 90 percent of Vernon County's shrinking base of pasture is not being managed adequately "to even get close to its capability for dairy and beef production." And largely due to the loss of pastured livestock, nearly 10 percent of crops are again being produced on erosive fields with over 20 percent slopes.

Clearly, grass holds the key to grassroots watershed restoration. As long as farms are diversified enough to have livestock needing pasture and hay, and as long as grain crops are raised in suitable proportion, on suitable lands, using suitable conservation tillage practices, there will be better retention of runoff and topsoil. Also, with comprehensive farm planning and improved management of crops and pastures, fewer nutrients and contaminants will enter streams and there will be less adverse impact to the receiving waters.

Aldo Leopold asserted that "a thing is right when it tends to preserve the integrity, stability and beauty of the biotic community." Leopold was never one to divorce aesthetics or ethics from the obvious need for more rigorous observation, recording, testing, and management. He believed in holistic integration and synthesis of knowledge, encompassing many disciplines and perspectives (Leopold 1949).

The small reservoirs of wildness and traditional life that remain on farms, the natural nooks and crannies that still survive in some of the world's more rugged and lovely landscapes, need to be protected now because their days are numbered. This is not just a matter of aesthetic preference, or even of ethical responsibility; it's an evolutionary necessity. Without reservoirs of biodiversity, how can ecosystems recover from overexploitation, cleanse themselves of contaminants, or adapt to future changes? The accelerating losses of biodiversity and family farms are the strongest indicators yet of declining land health. Maybe it will take another dust bowl to bring sustainable farming into the mainstream of agriculture. What a shame that would be.

References

Cohee, M. 1987. "Aldo Leopold's Conservation Legacy." Paper presented at Wisconsin Land Conservation Association Annual Conference, December 3, Oconomowoc, Wis.

Hawkins, A.S. 1940. "A Wildlife History of Faville Grove, Wisconsin." *Transactions of the Wisconsin Academy of Sciences Arts and Letters* XXXII: 29–65.

Helms, D. 1985. "The Civilian Conservation Corps: Demonstrating the Value of Soil Conservation." *Journal of Soil and Water Conservation* 40:184–188.

Hoops, R. 1988. *A River of Grain: The Evolution of Commercial Navigation on the Upper Mississippi River*. Research Report R3584. University of Wisconsin, College of Agriculture and Life Sciences. Madison.

Leopold, A. 1935. "Coon Valley: An Adventure in Cooperative Conservation." *American Forests* 41:1–4.

——. 1949. *A Sand County Almanac and Sketches Here and There*. Oxford University Press, New York.

Levins, R. A. 2000. *Willard Cochrane and the American Family Farm*. University of Nebraska Press, Lincoln.

Moore, R. 1998. *Refuge at the Crossroads: The State of the Upper Mississippi River*

National Wildlife and Fish Refuge. Special Report of the Izaak Walton League of America, edited by Zachary Hoskins. St Paul, Minn.

Thieling, C. 1998. "Important Milestones in the Human and Ecological History of the Upper Mississippi River System." Pp. 1–12 in *Ecological Status and Trends of the Upper Mississippi River System*, edited by K. Lubinski and C. Thieling. LTRM 99–T001. U.S. Geological Survey Upper Midwest Environmental Sciences Center, LaCrosse, Wis.

Trimble, S. W. 1999. "Decreased Rates of Alluvial Sediment Storage in the Coon Creek Basin, Wisconsin 1975–93." *Science* 285:1123–1125.

Trimble, S. W., and S. W. Lund. 1982. *Soil Conservation and the Reduction of Erosion and Sedimentation in the Coon Creek Basin, Wisconsin*. Prof. Paper 1234. U.S. Geological Survey. U.S. Government Printing Office, Washington, D.C.

Upper Mississippi River Basin Association. 1984. *Erosion in the Upper Mississippi River System: An Analysis of the Problem*. Upper Mississippi River Basin Association, Minneapolis, Minn.

U.S. Fish and Wildlife Service. 1987. *Master Plan for the Upper Mississippi River National Wildlife and Fish Refuge.* U.S. Fish and Wildlife Service, Fort Snelling, Minn.

Waters, T. F. 1995. *Sediment in Streams: Sources, Biological Effects and Control.* Monograph 7. American Fisheries Society, Bethesda, Md.

Chapter 5

Reading the Land Together

Wellington (Buddy) Huffaker

I am trying to teach you that this alphabet of "natural objects" (soils and rivers, birds and beasts) spells out a story, which he who runs may read—if he knows how. Once you learn to read the land, I have no fear of what you will do to it, or with it. And I know many pleasant things it will do to you.

—ALDO LEOPOLD, *River of the Mother of God and Other Essays by Aldo Leopold*

"I wanted to be a wildlife biologist, but everyone told me there weren't any jobs. So, I went into farming," claimed Duane Hohl, a landowner whose farm is about 5 miles as the crow flies from Aldo Leopold's Shack and farm along the Wisconsin River. To most people it would have been an innocuous comment made in passing. However, I received it more as testimony, revealing to me a great deal about the values Duane carries with him as he works his land. I also felt as if his words served as a bond that would exist even when we differed on our definitions of conservation. The moment crystallized for me the circumstances that had brought us together and, more importantly, the potential for collaboration.

We were in the middle of a volatile situation that characterized the frequently antagonistic relationship between agriculture and conservation. The U.S. Fish and Wildlife Service proposed to establish the Aldo Leopold National Wildlife Refuge in Sauk and Columbia Counties of Wisconsin. Because of the proximity to the Leopold Shack, which served as the setting and inspiration for much of his famous book, *A Sand County Almanac* (1949), the Service felt the refuge would be a tribute to his conservation legacy.

In the Draft Environmental Assessment, the U.S. Fish and Wildlife Service identified four approaches to conservation in their area of interest. The alternatives ranged from no action to its preferred alternative that encompassed a variety of land protection and conservation practices on 15,272 acres (USFWS 1999). At that point, the Service's preferred alternative sought to protect land primarily through fee title purchase from voluntary sellers but also included a role for easements. Its primary interest was in acquiring and restoring over 6,000 acres that were once an historic wetland complex. This area also included some of the most productive agricultural land in the two counties and is the site of Duane's dairy farm.

Just as time dulls one's memory of excruciating pain, it has allowed me to forget the community's reaction to the Fish and Wildlife Service's proposal. The local community responded to the refuge proposal by circling the wagons and preparing to defend their land and their livelihood. To Duane and many other local landowners, the wildlife refuge presented a range of threats, from pressure to sell their farms to extreme cases of crop depredation from increased wildlife populations. As the executive director of the Aldo Leopold Foundation, I was receiving a fair amount of heat from local landowners confusing the Foundation with the U.S. Fish and Wildlife Service. One local political cartoon insinuated that, in fact, I was behind the whole proposal.

Divisive land-use issues were already being hotly debated in the area. The Wisconsin Department of Transportation was proposing the expansion of a local highway from Madison to the Wisconsin Dells. Many local people felt the highway plan was an effort to appease interests in the Wisconsin Dells at the economic and ecological expense of the Baraboo area. In addition to this controversy, the U.S. Department of Defense announced it was going to "surplus" a 7,300-acre ammunition plant located 20 miles from the proposed refuge. The local landowners questioned why the federal government wanted to purchase 8,000 acres so close to where they were trying relinquish ownership of 7,300 acres.

Even more contentious was the conflict surrounding the change in regulations for the U.S. National Park Service's National Natural Landmarks. This Landmark status recognizes the Baraboo Hills as an ecologically significant geologic land form that covers 144,000 acres in Sauk and Columbia Counties and marks the southern boundary of the proposed refuge.

It was determined in 1999 that these regulations applied retroactively to the Baraboo Range National Natural Landmark established in 1980. The new regulations allowed landowners within a Landmark designation to petition for exemption from the status. Fueled by a private property rights coalition, numerous landowners mistakenly assumed that the natural landmark designation would limit landowner rights and allow the federal government to purchase land in the Baraboo Hills. Enough suspicion was generated so that 438 individuals out of approximately one thousand requested that their property be removed from the designation. This fear that the federal government was committed to taking over private property in the area certainly affected the public's receptivity to a federal wildlife refuge.

Several conservation professionals, myself included, warned the U.S. Fish and Wildlife Service that a wildlife refuge would probably not be well received. At a minimum, a great deal of footwork was necessary if the local community were to have a true understanding of the pros and cons of the refuge. The Service probably wished later that it had evaluated the political climate more closely. Perhaps it may have decided that the waters were already too rough to attempt such bold navigation.

The Service's plan was not without merit; the Lower Baraboo River area historically contained approximately 5,000 acres of contiguous wetland, including shallow emergent marsh, sedge meadow, wet prairie, floodplain forests, and tamarack swamps. This wetland complex was surrounded by uplands of prairie, savanna, and forests. As early as 1837, the first farms began to appear, and settlers found the basin was dominated by fertile soils and an abundant water supply.

Ironically, in the early 1900s the United States government, working with local farmers, began to channelize the existing historic waterways and construct drainage ditches, enabling the conversion of wetlands to cropland. Today, the area remains primarily agricultural and is home to a number of farming operations. In addition to supporting numerous dairies, the land also produces corn, soybeans, alfalfa, potatoes, and mint. The majority of the land is cropped or pastured every year. The area's topography, however, as well as fluctuations of the Baraboo River and its tributaries, make successful farming in some areas challenging. Remnant natural communities, areas where there has been very little disturbance from farming or other human land uses, can be found throughout the area. These areas include dry prairies, sedge meadows, oak forests, flood plain forests, and tamarack swamps. Varying in size from less than 1 acre to over 200 acres of oak forests, these remnants, in conjunction with farmland, provide valuable habitat for a variety of species.

As soon as the U.S. Fish and Wildlife Service began contacting landowners about its plan, the local news media wanted to run a story. Although the Service had only contacted a handful of the landowners in the area, it decided to release all of the information it had, including a map of the proposed refuge boundary. Many of the landowners were irate when they opened the local paper to learn their land was to be a wildlife refuge. They had not been consulted about their willingness to sell their property, and they saw the establishment of a refuge as a threat to their livelihood. But, I believe, the greatest offense to the landowners was the insinuation that after working the land—in some cases for generations—they were essentially told that they were not good enough stewards of the land. The farmers knew that the fields, riparian areas, and woodlands on their farms supported a great deal of wildlife while simultaneously producing food and timber, but they were being told that the federal government had to own the land in order for wildlife to thrive.

The scene was set for a drawn-out conflict and the potential deterioration of the working relationship that the Aldo Leopold Foundation had enjoyed with farmers. Fortunately, the Fish and Wildlife Service fiasco did not result in permanent community dissension. Quite the contrary; in fact, it created an opportunity for greater cooperation between farmers and conservationists. By giving farmers a voice in planning and decision making, they were able to protect and restore nature on farms. This chapter will explain how that process worked and the accomplishments that resulted. It illustrates what Leopold believed and what this book professes: that we can maintain natural habitat on working farms and that we must do so if we are to restore harmony between people and land.

Farming and Conservation Together

In the context of these conflicts, Duane's comment about wanting to be a wildlife biologist really gave me hope. It illustrated the fact that while the local landowners and the U.S. Fish and Wildlife Service appeared to be worlds apart from each other, there was some common ground. Duane voiced what many landowners were saying; they understood the importance of conservation but also knew that there are many ways to practice it.

In light of significant local opposition, the Fish and Wildlife Service recommended that landowners and concerned citizens develop an alternative plan for conservation. As I remember, several of us were drawn together by a mutual understanding of the potential for a constructive solution for both farming and conservation and, more importantly, a willingness to put our differences aside and focus on our shared values. We all agreed that although a positive relationship already existed between conservation and agriculture in the area, additional steps needed to be taken in order for this healthy relationship to continue.

The farmers acknowledged that some of the land being farmed was prone to flooding and might be appropriate for wetland restoration, while the conservationists acknowledged that agricultural land provided many benefits to wildlife and that measures could be taken to protect farmland. Both groups agreed that the biggest threat to wildlife in the area was not farming but urban sprawl. Subdivisions would provide wildlife with limited habitat, increase chemical applications on a per-acre basis, aggravate flooding through the increase of impervious surfaces, and present farmers with unsympathetic neighbors and other new challenges.

These shared values and an appreciation for the character of this landscape led to the founding of the Farming And Conservation Together (FACT) committee to develop an alternative plan. The voting members of the committee included three landowners, four representatives from affected local governments, and three conservation organizations. Nonvoting members included representatives from county, state, and federal agencies as well as a representative from the office of Wisconsin's U.S. senator, Herbert Kohl. Together we agreed upon a vision for the area: "a long-term balance of conservation and agriculture made possible through community-based oversight and coordination of voluntary enhancement options (USFWS 2001)." While I was taking a bathroom break at an early meeting, I was elected the committee's chair by consensus. All subsequent decisions were made only by consensus, and everyone was careful to use the restroom only during official breaks!

Initially, progress was slow. Mistrust led to tension among the group until ground rules were established and everyone became familiar with each other. With no model to follow, it felt at times as if we were moving backward rather than forward. In retrospect, I believe this was an important part of the trust-building process as we explored each other's motivations and perspectives. Relationships built on trust and understanding fostered honest dialogue. This dialogue resulted in a better appreciation of agriculture by the conservationists and better understanding of conservation by the farmers.

The most memorable step in the process was an open house hosted by Steve Luther, a committee member and farmer. All of the local landowners were invited, and the FACT committee cooked hamburgers and bratwurst for everyone. Nearly sixty people attended to listen to the committee's vision.

To be honest, there was such a large undercurrent of hostility when the process began that I could not have envisioned this assembly of people sitting in front of the community presenting a unified vision, nor could I have envisioned the warm reception that they received. At the open house, the audience went through a condensed process similar to that of the committee. They ranted and raved about the government's lack of respect and courtesy. But, slowly, the venting of frustration gave way to questions and comments about FACT's vision and plan, and by the end of the evening, the community had expressed its support and appreciation for the work of the committee.

The FACT committee developed a proposal that provides a framework for private landowners, conservation organizations, and governmental agencies to work together to pursue synergistic relationships between conservation and agriculture (USFWS 2001). In addition to promoting appropriate projects, the committee will serve as a forum where local citizens can present and examine ideas and concerns about agriculture and conservation.

Early in its deliberations, the FACT committee realized that even with an active corps of volunteers, development and implementation of the FACT vision required a dedicated staff position. The committee envisioned a coordinator that could serve as a liaison between the committee and the local landowners. In July of 2000, the committee received funding from the National Fish and Wildlife Foundation and hired Mike Jones as the FACT coordinator. Mike possessed the perfect background for the position. He had grown up on a farm in Iowa that his family still operates, earned a master's degree in wildlife ecology from the University of Wisconsin in Madison, and had practical experience working with private landowners through state agencies. Everyone agreed it was a good omen that someone of Mike's caliber and experience even existed.

The coordinator's job is to help landowners and conservation partners utilize programs and available resources to enhance the resource values of the area. Mike will provide technical assistance and help identify private sources of funding. One of the first things Mike did was to compile a matrix of current public and private agricultural and conservation programs available. He will connect landowners interested in certain practices with the proper agencies. As coordinator, Mike will also serve as the local contact point for landowners enrolled in the damage and abatement program. Landowners had expressed concern about the current amount of animal damage to cropland and frustration with the Wisconsin Damage Abatement and Claims Program and the potential for additional wildlife conflicts associated with future habitat restoration activities. Mike will assist landowners with administration of the program and with managed hunts intended to control wildlife populations.

Although the fundamental premise of the FACT proposal is to keep land in private ownership, landowners expressed a desire to provide additional benefits to the larger community. They will seek to create opportunities to educate the community on the relationship between agriculture and conservation. Demonstration and research projects will serve as living classrooms and laboratories for local schools, universities, agencies, and conservation organizations. They recognize that increased community participation in conservation practices would result in improved water quality and reduced flooding.

Funding for this initiative will come from public and private sources. The program matrix outlines existing voluntary programs available for many of the potential activities in our focus area. However, this funding will not be sufficient if FACT is to realize its vision. In order to address this limitation, FACT has already begun the process of requesting "Special Project" eligibility for the Wetlands Reserve Program and "Conservation Priority Area" status for the Conservation Reserve Program from the USDA Natural Resources Conservation Service. These programs provide substantial payments to landowners who undertake wetland or grassland restorations. Because of the broad-based citizen support for FACT's approach, the committee intends to ask its federal congressional delegation to assist it in seeking additional support from the U.S. Fish and Wildlife Service's private land program and Waterfowl Production Area programs. Private sources of funding also will play an important role. The support of the National Fish and Wildlife Foundation for the coordinator position reflects the interest that exists from private foundations and organizations. Finally, a number of landowners already actively practice conservation work on their land and are willing and able to cover the cost of the practices.

At the time of this writing, the FACT committee has not restored 1 acre of land or protected an acre of farmland from development. Many hurdles still exist before FACT can serve as a model for how conservation should be practiced. Landowners must fulfill their commitment to restoring marginal land, elected officials must secure funding for programs necessary to restore and protect land, and the conservation organizations and agencies must provide the resources and technical skills necessary to implement the vision.

There has been some criticism of FACT by those who view it as being a compromise for conservation. Some of these concerns are valid, and the process is slow, but I have no question that in the long run this approach will pay great dividends for both the land and the citizens of area. Already a sense of kinship and camaraderie has developed among the committee members that we hope will be transferred to the land.

These budding relationships have been expressed in various forms, several of which I have been the beneficiary. On one occasion, Steve Luther provided the Aldo Leopold Foundation's board of directors with a tour of his farming operation. Steve's willingness to visit with our board, and his complimentary remarks regarding the organization's role, affirmed that our efforts were improving our standing in the local community and furthering conservation. It was during this tour that I learned of Steve's connection to Leopold. One of his family's first plots of farmland was not far from the Leopold Shack, and after reading *A Sand County Almanac* in high school, he organized a field trip to the area. I was impressed to hear him talk about Leopold's writing and how he tries to balance the current economic demands of farming with stewardship.

On another occasion, when the Aldo Leopold Foundation was moving its office, Richard Gumz, a local farmer and committee member, brought a large flatbed trailer and two assistants to help us with the move. It turns out that Richard and I have in common the fact that we both graduated from Purdue University, as did his younger brother, Roderick, and that our degrees were all from the School of Agriculture (although our educational programs were different and we learned to read the landscape differently).

One spring afternoon while visiting the Gumz farm, I was delighted to see my first sandhill cranes of the year. Richard, Roderick, and I paused in the middle of our conversation to listen to their bugle. I quickly learned that the cranes' arrival was bittersweet to Richard and Roderick. They proceeded to tell me the exact date the first cranes arrived, which crops cranes liked best, how much damage they do on an annual basis, and the location of many of their favorite nesting spots. I realized that the brothers studied the cranes' behavior as ecologists would and were actually more observant than many ecologists I know.

Conservation Work of the Aldo Leopold Foundation

Because of the positive personal and professional relationships I have gained from this experience, I find myself using it as a measure all of the activities of the foundation I oversee. Established by the family of Aldo Leopold, the Aldo Leopold Foundation seeks to foster a society that has a conservation ethic at the core of its values and actions. We pursue our mission, which is "promoting harmony between people and land," through programs in environmental education, land stewardship, and ecological research.

We believe that an informed, thoughtful, and ethically oriented society is necessary if we are to ever realize our goal of a widely embraced conservation ethic. Our educational activities focus on increasing the general public's appreciation and awareness of our conservation heritage through tours of the Leopold farm, seminars on conservation issues, talks about Aldo Leopold and his work. Technical workshops on issues ranging from prescribed burning to wetland restoration provide landowners the information and skills necessary to express a conservation ethic on their own property.

However, Leopold knew that actions speak louder than words, and our land stewardship program guides us to demonstrate good stewardship as well as to talk about it. The cornerstone of our activities is the original Leopold farm. We continue to improve the health of the land through a variety of restoration and management activities. Pine plantations, duck ponds, and prairie restorations tell their own story of how conservation has evolved since the 1930s.

But Leopold did not work to restore land health only on his property. While a professor of wildlife ecology at the University of Wisconsin, he was instrumental in setting up community-based conservation initiatives throughout the Midwest such as the Coon Valley Erosion Project (see chapter 4). Following his lead, the Aldo Leopold Foundation seeks to develop community-based solutions to environmental issues throughout south-central Wisconsin. There is a strong commitment to engage and involve neighbors, landowners, and volunteers in the projects. This approach allows us to play a critical role in improving land health on more than 15,000 acres through partnerships with over twenty private landowners, thirty nongovernmental organizations, and five government agencies.

One such initiative grew out of a atmosphere different than that of FACT. For the past ten years, utilizing the collective abilities of landowners, a trained ecologist, and an army of volunteers, The Blufflands Project has been maintaining and expanding vanishing savannas and prairies in south-central Wisconsin.

The Blufflands Project was borne more out of curiosity than of conflict. Jeb Barzen, a self-proclaimed "duck biologist" and director of field ecology at the International Crane Foundation (and FACT member), began by investigating oak savanna restoration on his property. When Jeb burned the savanna on his land, he had an opportunity to engage his neighbors. After watching Jeb conduct a safe prescribed burn, they began to ask him why he was burning at all.

Jeb, the prairies, and the savannas had been waiting for just such an opportunity. Since European settlement in the mid-1800s, agriculture, human development, and a cessation of fire have resulted in the loss of all but one-half of one percent of Wisconsin's original 2.1 million acres of grassland and all but 500 acres (one one-hundredth of a percent) of the 5.5 million acres of oak savanna (WIDNR 1995). Today, many of these remnants are scattered on the beautiful but rugged bluffs of the Wisconsin River, most no larger than 2 acres, providing a tenuous link to the communities that thrived in this area for 6,500 years before European settlement (Winkler 1985).

Over time, Jeb helped others in the area understand and appreciate these ecological gems. And it was not long before landowners began to request assistance in maintaining and restoring remnants on their land. Jeb often says that these individuals were ready and willing to "do the right thing" but did not know what to do or how to get started. In his spare time, and with the help of a cadre of volunteers, Jeb gradually increased the trust between himself and his neighbors, and, consequently, was able to influence the acreage under their management.

Eventually, it became apparent that there was far greater demand from landowners interested in management and restoration than Jeb and the volunteers could provide. The initiative has evolved into a more cohesive project that is cosponsored by ALF and the Prairie Enthusiasts, a not-for-profit organization dedicated to preserving prairie and oak savanna remnants in the upper Midwest. This added organizational infrastructure has allowed us to hire a halftime ecologist to help refine and further develop The Blufflands Project.

Project sites are selected based on their ability to sustain and expand remnant prairies and savannas in the Lower Wisconsin River Valley of south-central Wisconsin. Among the rarest plant communities in the world, oak savannas and prairies support numerous endangered, threatened, or species of special concern, eleven of which are found on The Blufflands Project sites (Lange 1998; WIDNR 1994a, 1994b). We estimate that without active management for as little as ten to twenty years, many of the remaining remnants could be lost forever. Landowners interested in participating in the project are provided with the professional expertise and labor necessary to implement a variety of land management techniques that can reverse this trend.

Volunteers and staff (and with Jeb's continued participation) have developed management plans, conducted prescribed burns, controlled invasive species, restored prairies, and inventoried vegetation. Currently, the project assists nearly forty landowners possessing over 3,000 acres. Direct management occurs on over 1,000 acres at this time. I must stress the importance of the voluntary nature of the project. Although some landowners receive cost-sharing from state and federal programs for certain projects, the money usually is used to help subsidize the costs of materials. The incentive for the landowner is strictly the personal satisfaction derived from being a good steward of the land.

Many of the project sites are also adjacent to land owned and managed by the Wisconsin Department of Natural Resources, creating large, contiguous tracts of managed land. An example of this is Matt Millen's 700 acres encompassing nearly a mile of waterfront and four remnant bluff prairies along the Wisconsin River. It is located next to nearly 500 acres of land owned by the Department of Natural Resources. Both Matt and the Department of Natural Resources have been restoring unproductive agricultural fields to prairie and removing windrows to increase the overall size of these grasslands. These measures provide substantial benefits to a suite of grassland bird populations (e.g., meadowlarks, bobolinks, and upland sand pipers) on the decline in the area because of the disappearance of large grassland acreage necessary for breeding territories (Sample and Mossman 1997).

Although our management efforts have purposefully targeted adjacent properties to increase our effectiveness at maintaining the ecological value of these remnants, they still exist in a mosaic of land use that includes agriculture, recreation, and development. Several of the landowners in the project continue to farm portions of their property more suitable to row crops, demonstrating again that agriculture and conservation need not be mutually exclusive.

Another component of The Blufflands Project is to examine research questions designed to benefit both the private landowner and land managers of prairies and savannas. We are interested in the feasibility of generating income from working lands by restoring the land to prairie. The high price of prairie seed combined with the growing demand may allow landowners to generate income from the sale of seed to local nurseries or other landowners.

Responsibilities and Rewards of Conservation

Recently, Steve Swenson, the Aldo Leopold Foundation's ecologist, was giving a presentation about The Blufflands Project and he shared a story that exemplifies its importance. Steve had spent the day with Bob Taylor, a landowner, walking Bob's 80 acres. This incredible prairie remnant has a breathtaking view of the Wisconsin River valley only 30 miles down river from where Aldo Leopold began restoring prairie over sixty-five years ago on his own farm. The property represents the best and worst of what is now typical of these rare remnants. The mature oaks interspersed among the rich prairie flora are being displaced by encroaching red cedars.

Over the course of the day, Bob pointed out to Steve a new clone of sunflowers, the location of bull snake hibernaculums, cedars that needed to be removed so that remnant prairie vegetation would not be overshadowed, and other management issues needing attention. He was, in effect, reading his landscape aloud to Steve, so that they could work together to develop a strategy to improve the health of the land.

I think Bob probably could have done this before The Blufflands Project was established. He is like many landowners who understand a great deal about their land and observe the subtle nuances of change over the seasons. They pay attention to where the wildlife congregate and where and when to see the best wildflowers. However, Bob's participation has increased his understanding of how important the role of management is, and it reinforced his appreciation that his property is a valuable ecological treasure. He developed this appreciation on his own over many years by walking the land and gazing out over the valley from the top of his bluff. But it had no outlet.

Bob now shares his interest and commitment to stewardship with other landowners and has the resources of trained ecologists and volunteers to assist him in achieving his management goals. The concept of placing a conservation easement on his property has even been discussed as a way to ensure the work he is doing will be honored by future generations. He has come to better realize not only the rewards of good stewardship, but also the responsibilities. Aldo Leopold once described conservation as occurring "when the land does well for its owner, and the owner does well by his land; when both end up better by reason of their partnership" (Leopold 1991). Surely, this description fits the relationship between Mr. Taylor and his property.

Aldo Leopold also wrote "In our attempt to make conservation easy we have made it trivial" (Leopold 1949). He warned us not to assign too much responsibility to public agencies. Yet I am afraid that we have fallen into a similar trap by assigning the work of conservation to professionals (public and private) and trivializing the role of the landowner. This might be an acceptable outcome if conservation were actually furthered in this scenario. But when we look around, we see that our landscape is changing and, in most cases, to the detriment of our native flora and fauna.

There are no easy solutions. We all read the landscape based on our own experiences and values. By broadening the conservation discussion, we get a richer and fuller picture of what the land means to our communities and what it can be in the future. One need look no further than the results of FACT and The Blufflands Project to see the pleasures available to us when we take the time to learn together.

Acknowledgments

A special thanks goes to the members of FACT and The Blufflands participants for their involvement and willingness to share their stories. The author is also indebted to the following for their review and comments on this chapter: Marcy Huffaker, Rob Nelson, Nina Leopold Bradley, Steve Swenson, Mike Jones, Jeb Barzen, and Duane Hohl.

References

Lange, K. I. 1998. *Flora of Sauk County and Caledonia Township, Columbia County, South Central Wisconsin.* Technical Bulletin No. 190. Wisconsin Department of Natural Resources, Madison.

Leopold, A. 1949. *A Sand County Almanac.* Oxford University Press, New York.

———. 1991. "The Farmer as a Conservationist." Pp. 255–265 in *River of the Mother of God and Other Essays* by Aldo Leopold, edited by S. L. Flader and J. B. Callicott. University of Wisconsin Press, Madison.

Sample, D., and M. Mossman. 1997. *Managing Habitat for Grassland Birds: A Guide for Wisconsin.* Wisconsin Department of Natural Resources, Madison.

U.S. Fish and Wildlife Service (USFWS). 1999. *Proposed Aldo Leopold National Wildlife Refuge: Draft Environmental Assessment and Interim Comprehensive Conservation Plan.* U.S. Fish and Wildlife Service, Fort Snelling, Minn.

———. 2001. *Proposed Aldo Leopold National Wildlife Refuge: Final Assessment.* U.S. Fish and Wildlife Service, Fort Snelling, Minn.

Winkler, M. 1985. "Late Glacial and Holocene Environmental History of South-Central Wisconsin." Ph.D. diss., Institute of Environmental Studies, University of Wisconsin, Madison.

Wisconsin Department of Natural Resources (WIDNR). 1994a. *Rare, Threatened and Endangered Species and Natural Communities in Dane County.* Publication ER-213. Wisconsin Natural Heritage Inventory, Bureau of Endangered Resources, Madison.

———. 1994b. *Rare, Threatened and Endangered Species and Natural Communities in Sauk County.* Publication ER-256.Wisconsin Natural Heritage Inventory, Bureau of Endangered Resources, Madison.

———. 1995. *Wisconsin's Biodiversity as a Management Issue: A Report to Department of Natural Resources Managers.* Pub-RS-915 95. Wisconsin Department of Natural Resources, Madison.

Chapter 6

When Farmers Shut Off the Machinery

Brian A. DeVore

The Canada thistles on David Podoll's farm aren't looking so good these days. These prickly weeds are being wracked by one disease that leaves a "rust" on the plant and another that attacks the roots. To top it off, the painted lady butterfly likes to lay its eggs in the thistle's flowering head. When the larvae hatch, they munch their way out, pretty much shredding what's left of the already ravaged host.

The result is a natural weed control system that no amount of spraying or mowing could accomplish. So what does the southeast North Dakota farmer do when he sees Canada thistle plants in the uncultivated margins next to his 480 acres of small grains and pasture?

He leaves them.

Podoll explained to me that his thistle control program relies on a natural repository of disease and weeds. One grand, expensive weed eradication campaign might eliminate all the thistles temporarily, but it would also destroy a natural source of future control. Thistle rust and the painted lady butterfly need places to overwinter. Destroying their host would be like lighting a match to a formula for a really good weed killer. That's why Podoll purposely protects and encourages the establishment of ecological "edge" areas next to his fields—stands of trees, soggy sloughs, natural grass areas—that are not disturbed and thus provide a place for the Canada thistle, and its enemies, to spend the snowy months. The added benefit of this strategy is a greater diversity of wildlife, insects, and plants on the farm: things that are important to the farmer.

I must say, I was mesmerized when Podoll first told me this tale of how one can make a thistle's life pretty miserable without the help of Monsanto or DuPont. It made such perfect sense in a clear yet sophisticated way. How,

I asked the farmer, did you ever figure out this fine balance between weed control and ecological protection?

Podoll's answer was simple: "You just watch."

The Eyes Have It

That's a significant statement. Few of Podoll's farming contemporaries even cast a sidelong glance at the ecological workings of their fields, let alone take the time to "watch." I've talked to extension educators who recommend applications of nitrogen fertilizer or a certain type of pesticide, regardless of the state of the soil's fertility or how many plant-eating bugs are present. Drugs are as much a part of a confined farm animal's diet as corn and soymeal. Many farmers consider the money forked over for these potentially wasteful applications "insurance" well spent in these days of relatively cheap agrochemical inputs. Farmers, scrambling to manage more acres and bigger livestock herds—often while working off the farm to make ends meet—take such advice and run. Such a harried way of managing makes it difficult to notice if one "solution" is creating many other problems that require even more stopgap measures.

But Podoll and other farmers attempting to raise food and fiber in a more environmentally and economically sustainable manner are learning that often the most valuable assistance is not available through extension agents, how-to manuals, input suppliers, or any of the other conventional sources of agricultural production information. Instead, they are increasingly relying on their own senses to gather the data needed to make sustainable farming decisions. Physically, these farmers may not be any more eagle-eyed than their neighbors. In a sense, it may be what these sustainable farmers *lack* that sets them apart. One top-rate sustainable farmer and self-described land watcher says the key behind good observation is to look at something without allowing preconceived notions to interfere. It's not just looking at the land in a different light, it's looking at it in a *clear* light unadulterated by attitudes that what worked in the past will work again, or that the next "silver-bullet solution" is just around the corner.

What many of these farmers are discovering is that the next step after close observation is accepting that natural processes work best when disturbed as little as possible. The result can be farms that are ecological treasures, as well as generators of a good standard of living. And that is what this book is all about.

Over the years, I've talked to farmers representing a myriad of geographical locations and farming systems who have stories akin to David Podoll's—examples of how getting off the tractor, shutting off the engine, and walking the fields can illuminate nature-based solutions to problems that conventional farming tries to bludgeon into submission with chemicals, machinery, drugs, and all the other trappings of industrial agriculture. Ecologically sustainable farms of the future will be populated with people who have open eyes—and open minds. This chapter is about some of these farmers and how they used their observation skills not only to confirm what they already knew about their interactions with the land, but also to take them into new and exciting agroecological territory.

Battle Reports

Some of the most fascinating "monitoring stories" are those that shed new light on the weed and insect pests that can make the lives of farmers so difficult. After all, it's these struggles between farmer and pest that create the most potential for bringing agriculture and nature into conflict. Podoll's experience with thistles is a good example of how that struggle does not have to end badly for one of the parties involved. Western Oregon salad greens producer Frank Morton has had a similar experience, albeit in a much different environment than the northern plains. His entomological epiphany came while snapping some "pretty garden photos" one day. One of the pictures Morton took was of a bald faced hornet dangling from a plant by its hind legs, chewing on the head of a fly; now *there's* an image that sticks with a person. But Morton used it as more than gruesome storytelling fodder. He went on to note the overall toll hornets and yellow jackets were taking on aphids, flies, and caterpillars, the bane of gardeners. Through further observation, he realized that for every, say, seven insects he'd see on a plant, four of them were there simply to feed on the three pest species. This fits well with Morton's farming philosophy, which is based on a Japanese wild gardening concept that works with natural processes as much as possible. Morton estimates that there are two hundred species of plants in the garden at any one time; he and his wife Karen planted perhaps half of those themselves.

"Now our attitude is, if you don't know what it is, you don't kill it, you watch it," Morton told me. "Any gardening book is going to tell you if you have a plant covered by aphids, yank it out and put it in the compost pile. That's not what I do. I look for what's attracted to the aphids."

What the Mortons have discovered on the highly productive 2 acres they farm is that in a balanced system, beneficial insects will keep the pests in check. But such a balance is not easy to obtain. It requires plantings that attract beneficial insects as well as leaving naturally occurring habitat alone. And perhaps most importantly, it means realizing that no pests at all may be bad news for long-term control of their populations as the beneficial insects run out of food and experience a population crash.

"You must have the pest to have the predator," said Morton. "I get nervous if I don't see any pests."

These kinds of stories are important for a couple of reasons. For one, they serve as inspiration for ecologically minded farmers who are toiling within an economic, political, and social environment that narrowly defines success in terms of bin-busting yields and does not recognize or reward stewardship practices. In addition, these reports from the field are providing practical, sustainable guidelines for what works day in and day out on our nation's farms.

Finally, certain kinds of monitoring is reminding farm families that there is more to quality of life than a simple correlation to "standard of living." Some have even made personal sacrifices for the sake of key biological indicators. Art Thicke, a southeast Minnesota dairy farmer who has become an avid birder in recent years, no longer uses his four-wheeler to take cattle to and from the pastures during the spring and summer; the vehicle's noise inhibits his ability to observe the birds. That's a dramatic shift in habit for the former stock car racer.

"With the four-wheeler we used to just race out and open and close the gates and then race back in," Thicke said one summer afternoon as he watched a pair of bobolinks do their black-and-white flutter in a stand of grass. "Now I'm enjoying the whole experience. When I see something like a scarlet tanager, it makes my day."

Star Power

Now contrast the image of Art Thicke's head jerking skyward as a tanager passes overhead with the one I witnessed in a darkened conference room on a June afternoon in 1996. On that day I sat amongst agronomists, engineers, and others participating in something called the International Precision Agriculture Conference and watched a video of a John Deere 7800 tractor rolling along a California test plot. The man sitting in the tractor's seat threw up his hands to show he wasn't driving. Instead, four global positioning satellite (GPS) receivers mounted on top of the cab were using signals from space to guide the machine's journey back and forth across the field.

"We think this is the first step toward totally automatic farm implements," crowed a Stanford researcher while the tape was rolling. Indeed, while thumbing through a farm magazine four years later, I spied a photo of a John Deere tractor pulling a sprayer. It caught my attention because there was no seat or steering wheel on the tractor—it was a prototype of a GPS-guided driverless field implement.

Such technology proves that the importance of knowing the land intimately has not been lost on the boosters of industrial agriculture. But they also know that firsthand monitoring puts a size limitation on farms and makes operation of large, complicated equipment a hindrance. Their response has been the development of technologies that break the land down into electronic bits and bytes. GPS equipment, wedded with yield monitors, field maps, and all the other high-tech gizmos that fall under the "precision agriculture" umbrella, can now tell a farmer that 1 square yard of a field is producing fewer bushels of corn per acre than another square yard. The farmer can then use that information to apply more (or less) fertilizer here, or to justify the cost of draining a piece of soggy ground there. Precision agriculture's supporters say the technology will make row-crop production more environmentally sustainable because it will lead to more precise applications of fertilizers and other inputs, thus reducing runoff and saving energy.

It's been said an innovator has to travel 50 miles before anyone will listen to his or her "expert" advice. Well, the idea behind GPS technology seems to be that information from the land has to travel to the stratosphere and back before it's considered worth paying attention to. And once that information is broken down into streams of electrical pulses that computers can read, why not skip the middleman and just allow Pentium Chip plow boys to put that information to work?

Perhaps there will always be a need for farmers who are familiar enough with the land to interpret the piles of maps all that software can produce. Indeed, it's doubtful the rural landscape will be crawling with robotic John Deeres operated from a Chicago office tower anytime soon. But in other ways, precision farming could finally clear the way for making crop farming even more like a mindless, farmerless, manufacturing enterprise.

After watching the video of the satellite-guided tractor, I telephoned David Podoll and asked if he thought such "precision farming" could ever bring about a more environmentally benign form of agriculture. To him, the question was a no-brainer. For one thing, the data being fed into those high-powered computers was pretty superficial from a farmer's point of view. A satellite couldn't provide insights into the intricate relationships among Podoll's thistles, a rust disease, and a species of butterfly. Knowing the corn yield and maybe even the amount of fertility contained in a square yard of soil is not the same as knowing such characteristics as the amount of healthy bacterial activity that's taking place. Podoll's final word on this technology was that before computers spit out even more "answers" about agriculture, precision farming's practitioners need to take a step back and ask a basic question: "Are we caring for the land better or are we just using precision farming techniques to grow what we want the land to produce?"

The Real Questions

True on-farm monitoring isn't just some sort of nice benign tool for cutting chemical usage here and saving a little soil there, providing an agronomic Band-Aid so that crops like corn and soybeans can be raised a few more years down the road. Sustainable farmers use their observations to ask just the type of questions Podoll calls for. More precise applications of fertilizer in a corn field may reduce water contamination and save some tractor-riding time, but what if that field were converted to a production system that didn't require chemicals or John Deeres? Draining that wet spot in the back forty is bad for wood ducks but good for soybeans. So how does the welfare of the farm family rate in this equation?

As Virginia farmer and motivational speaker Joel Salatin once told me: "We have become incredibly accurate at hitting the wrong target."

But there are some excellent examples out there of how observation of one aspect of a farming operation can prompt a farmer to make some key connections and start hitting the right "target." David Podoll could have reacted to the observation that rust kills Canada thistle by trying to figure out how he could introduce the disease onto the plants artificially. But he saw the ecological chain that links this disease with his natural areas on the farm. The path he took accomplished many goals at once, rather than just dealing with a single, isolated problem as if it had no connection to anything else on the farm.

For many farmers, one of the first pieces of the ecosystem they closely observe on their land is the soil. Rampant erosion or a field that tests low in nutrients and organic matter is a sign to the farmer that production changes are in order. Some react by building terraces to keep the soil in place or pouring on more nitrogen to maintain fertility levels.

Then there are farmers like Mike and Jennifer Rupprecht. These crop and livestock producers farm in a rugged, erosion-prone area of southeast Minnesota. They have always taken steps to keep their soil on the land and have gained a reputation throughout the region for putting stewardship ahead of bottom-line goals. So it was a real eye-opening experience for the farm family when, in 1994, a soil sample unearthed a piece of unpleasant history: 48 inches of eroded topsoil lay at the bottom of one of their draws. Would the controlled rotational grazing system and conservation tillage methods the Rupprechts had adapted a few years before prevent such erosion and improve soil quality?

Through observation, the Rupprechts concluded that the presence of worms and other biological activity in their soil says a lot ultimately about their farm's profitability, environmental sustainability, and even the family's quality of life. Good, healthy soil means better yields. But it also means more plant diversity. And, as the Rupprechts eventually discovered, a vibrant, biologically rich landscape results in more wildlife on the farm, cleaner water, and a more pleasant work environment. I've visited their farm many times, and it's clear the Rupprechts now see their farm in a big-picture, holistic sort of way. They also know that certain indicators, such as nesting grassland songbirds, can tell them just as much about how they're doing as farmers as the deepest soil probe.

"We kind of came at it from the soils perspective and then discovered birds and went from there," recalled Mike recently. "That was exciting."

The Monitoring Team

The Rupprechts came to their realization partially through their involvement with a monitoring project sponsored by the Land Stewardship Project. As is described in chapter 1, this was a concerted effort to make on-farm observation an integral part of managing a sustainable operation. It also was successful in showing that once farmers make monitoring just one ecological indicator part of their routine, they can have their eyes opened to all sorts of connections on their operations. Making those connections can prompt the kinds of changes that are just not possible through regulation or extension service field days.

Consider the impact bird monitoring had on the six Monitoring Team farm families, all of whom were experimenting with a grass-based sustainable livestock production technique called *management intensive rotational grazing*. One summer afternoon, I sat in the back of an aging Chevy Suburban truck as it bounced along a fence line, its occupants blurting out the names of birds flitting about in a nearby pasture.

"Flycatcher!"

"Grackles!"

"Eastern kingbirds!"

"What's that? A savannah sparrow?"

"No, it's a fence tightener," announced one of the birders with a laugh after a quick check with the binoculars.

This wasn't a group of urban ornithologists talking excitedly about the difference between a songbird and a hand-sized piece of ratcheted steel. This time, farmers were the ones packing the binoculars and field guides on a tour of the Brian and Carol Schultz farm in south-central Minnesota. They were being given a mini-course on the feathered residents by Art "Tex" Hawkins (see chapter 4), a U.S. Fish and Wildlife Service biologist, and Art Thicke, the dairy farmer who has made birding a part of his livestock chore routine.

Both men are founding members of the Monitoring Team. Their enthusiasm for birds is infectious, and it's easy to see why birding became one of the Team's most popular monitoring tools. It's user friendly *and* can put the development of sustainable management techniques in the hands of the farmers themselves. Because it can be worked into livestock chores like moving cattle and fixing fence, birding is a handy way for gauging some of the impacts a farmer is having on the land. It's also more pleasant than grubbing up soil samples.

"I think of all the monitoring tools, birding is the most fun," said Thicke. "It's addictive."

How certain monitoring activities rate on the "fun scale" while fitting into daily farm activities is not trivial. During the mid-1990s, John Doran, a USDA soil scientist based at the University of Nebraska, tried to develop a comprehensive soil quality testing "kit" for farmers. The kit contained resources for testing, among other things, soil respiration, infiltration capacity, bulk density, acidity levels, nitrate levels, electric conductivity, and compaction. Taken together, all of these indicators should give farmers an excellent idea of how their farming practices are affecting the biological health of the soil. But when, as a dry run, Doran sent the kit out to a few innovative farmers who were trying various sustainable practices, he was disappointed in how they used it. The farmers reported back that they simply didn't have time to fit the kit into their routine, Doran told me.

Some farmers were overwhelmed by all the tests that could be done, and simply picked and chose indicators at random as if they were grabbing different-sized wrenches out of a tool box. Such selective use of the kit may have helped the farmers determine, for example, if their soils were short on nitrogen or too acidic, but it didn't give a general overall picture of soil quality. That's why one no-till farmer in Illinois was able to use the kit to reaffirm his belief that his intensive use of chemicals was good for the soil. Doran's experience with the soil quality testing kit reinforced his belief that to be truly useful for farmers and scientists, monitoring systems must give qualitative measures—sights, sounds, and smells—the same weight as their quantitative counterparts—hard numbers on pH and nutrient levels, and the like.

Dickcissels and vesper sparrows aren't the end-all indicators of how a farm is doing ecologically, but their presence or absence tells a big picture story that's hard to fudge. Wildlife biologists consider grassland bird species to be good biological barometers in farm country because they respond so quickly to changes in land use—bad and good, according to Hawkins.

That's why it's so exciting that by the third year of the monitoring initiative, bird sightings and activities were among the first items farmers mentioned during their monthly reports. When they got together for meetings or field days, farmers were not bragging about their corn yields, milk production, or even improvements in soil structure. But they were quick to let each other know about the number of successful bluebird or bobolink nestings they had witnessed on the land.

Hawkins spent time walking the fields with the farmers. Hawkins—the kind of guy who's not afraid to stand in front of a group of livestock farmers and make "chewy-chewy" noises with his mouth to demonstrate a bird call—also trained several local Audubon Society members with excellent birding skills to assist the farm families with initial counts.

"The bird that first hooked me was the eastern kingbird," recalled Mike Rupprecht. "I would see them by the cows, darting after flies. It was quite entertaining. We sure enjoy the show of them chasing other birds."

When their birding skills grew, the farmers, with the help of some local bird enthusiasts, were able to identify and record some of the more threatened grassland nesting birds such as meadowlarks, bobolinks, and dickcissels, as well as vesper and grasshopper sparrows.

As they became more aware of the bird life on their farms (and began enjoying the birds more), the farmers took the next step: they started wondering what impacts their livelihoods were having on the life cycles of their feathered neighbors. At team meetings, the farmers began discussing with Hawkins concerns they had about nesting disruptions caused by haying, pasture clipping (a method for keeping the grass more palatable for livestock), and even grazing. Although several of the farmers observed that cattle were sometimes able to graze the paddock with an active nest in it without destroying the nest eggs or nestlings, the results weren't always as positive when it came to mechanical forage harvesting.

"One time when I cut hay on our other farm I must have destroyed five to ten batches of bobolinks," recalled Thicke. "The adults were all circling, screaming at me."

Changing Farming Practices

It became clear the hay fields and managed pastures were in danger of becoming avian "population sinks," or booby traps, rather than "population sources." As a result, several farmers reduced or delayed pasture clipping to allow for fledglings to achieve some level of mobility before the mower disrupted the nest. Mike Rupprecht did not clip any of his paddocks one year. It didn't appear to have any negative effect on the productivity of the pastures or beef cow herd, and the Rupprechts observed a number of male dickcissels using taller plants in their pastures as singing perches. Thicke also eliminated clipping on some of his paddocks one year and by the fourth grazing, he said, "You couldn't tell where you clipped and where you didn't." In other words, despite the lack of clipping, the cows still found the grass palatable.

One of the management techniques that has the farmers most excited about improving grassland species nesting success is the establishment of "rest areas" within their managed grazing paddock systems. This is the grass farmer's version of leaving a piece of land idle for part of a season, allowing the vegetation to grow undisturbed by grazing. In 1995, each farm held one paddock out of grazing from the beginning of the season until at least the end of July (most farmers on the Monitoring Team had twenty to thirty fenced paddocks).

The densely vegetated rest areas provide a place for birds to nest undisturbed by cattle or machines. They also allow birds disturbed in adjoining paddocks to retreat to the lush cover and re-nest. Farmers noticed greater concentrations of bobolinks and dickcissels in the rest areas the first year of the experiment. A search of a rest paddock on the Rupprecht farm in 1996 confirmed a successful bobolink re-nesting. By the end of the nesting season in late July of that year, Art Thicke and his wife Jean saw more than sixty bobolinks, some of which were fledglings, flocking together on their farm.

In addition, the rested paddocks give grass and legume seeds an opportunity to mature so they can re-seed either directly or through the livestock. Allowing the grasses and legumes to grow for a longer period also increases the root structure of the plants, thus improving soil structure. The term "win-win" is much used and abused these days, but that's what we seem to have here. Since these rest areas are showing a benefit not only for wildlife but also for pasture productivity, these are management practices that benefit the farm financially, *as well as* improve the environment.

The relationship that has developed between farmers and birds in southeast Minnesota is a prime example of how monitoring can help a farmer overcome the belief that profits and ecological health are mutually exclusive. Such close attention to the land's details can also soften hard-and-fast rules about what's good for the environment.

The Stream Team

One brilliant fall day, I found myself stumbling after Larry Gates and Ralph Lentz as they hiked along a short section of Sugarloaf Creek in southeast Minnesota. This section of the stream had grass-covered banks with a gentle slope to them. The channel was deep and there were overhanging areas at the water's edge, perfect habitat for fish and other creek residents. The two men headed downstream and Gates kicked up a leopard frog, an increasingly rare sight in farm country. Then a tiny shrew tore itself free of the overhanging grass and dropped into the creek. In a burst of panicky energy, it motored the few feet of water to the other side and scrambled up the grass-covered bank.

To put an exclamation point on the stream's already excellent bill of health, Gates squatted next to the fast-running water and scooped up a handful of the creek bed. He cracked a smile as the water drained through his fingers, leaving a mound of clean gravel. The presence of relatively silt-free alluvial material was a sign that little erosion was coming off the pastures adjacent to the creek. It was also an indication that the current was running fast enough to cleanse itself of excess silt.

A creek doing this well can appeal to all the senses. Lentz stepped into the channel and cold water slurped around his rubber chore boots. The stream was making the kind of "babbling brook" sounds associated with fast, narrow waterways.

"I like the sound," Lentz said as he waded against the current. Gates agreed: "It's turning into a gurgler."

This is a far cry from the kind of waterways normally found in farm country: wide, sloppy creeks filled to the brim with chocolate braids of silt-carrying water. Constantly sloughing cliff-like banks devoid of vegetation make it almost impossible to stroll down to the channel for a closer look at the state of things.

Sugarloaf Creek's good health is due in large part to a strategy that utilizes cattle—long considered the enemies of healthy watersheds—to improve the stability of the stream bank. Gates is a Minnesota Department of Natural Resources (DNR) watershed coordinator. Lentz is a farmer. Both are founding members of the Monitoring Team. The stretch of Sugarloaf that looked so good on this particular day winds through Lentz's 160 acres before flowing another three miles to the Mississippi River.

More than three decades ago Lentz approached technicians in the local Soil Conservation Service office (now the Natural Resources Conservation Service) about creating a conservation plan for his portion of Sugarloaf Creek. What they suggested was the standard recommendation of the time: fence the stream off, plant trees and, most of all, keep the cattle out. There's sound reasoning behind such advice. Several studies throughout the years have shown significant environmental benefits from creating protected riparian areas along stream corridors (Barling and Moore 1994). So in 1967 Lentz fenced off 4 acres along the stream, planted spruce, pine, cedar, and white ash, and sat back to watch what would happen, convinced he had done the right thing. In fact, the farmer's initial plan was to fence off the entire creek where it ran through his property, creating a permanent riparian strip along both sides of the stream. He grazes approximately 100 acres and doesn't really need the forage found next to the stream to make his cow-calf operation pencil out economically.

But things get busy on the farm and Lentz never got around to building more fence. And it was beginning to look like maintaining a permanent riparian area was a lot of work, anyway. Over the years, a couple of major floods wrecked the posts and wire, forcing the farmer to rebuild and perform more maintenance on the fencing than he would have liked.

Lentz didn't know this in 1967, but by not getting around to fencing off the whole stream, he had created a perfect laboratory for comparing different land uses on a stretch of creek roughly a quarter of a mile long. By 1989, his makeshift demonstration plot began telling an interesting tale.

I've been to the Lentz farm several times over the years and never fail to be amazed at the contrasts present on this quarter-acre stretch of Sugarloaf.

The fenced-off area, now heavily forested, is host to a wide, shallow stream with erosion-prone banks. It's almost impossible to wade through the sunless stretch because of the mucked up bottom. The trees have grown so well that they've shaded out the grasses and other undergrowth that hold soil together.

And, as I described before, the section right above the fenced-off area, where Lentz retarded succession by allowing cattle to periodically graze, is a scene right out of a trout angler's dream. It's difficult to overstate just how much the creek changes just within a matter of a few yards.

"I was very surprised to see the fenced off area deteriorating," recalled Lentz recently. "What I had been taught was not what I was seeing."

He became convinced that simply planting trees along a stream bank was not the answer. In fact, Lentz began to believe that in some cases allowing cattle to graze along a stream on a limited basis could improve the waterway considerably by opening it up to more diversity of plant life.

The farmer had a hard time getting people—especially natural resource professionals—to listen to him. It wasn't like he was claiming that trees were bad for stream banks and livestock should be allowed to run amok in our floodplains. Lentz just wanted people to take a second look at the "creeks and cattle never mix" mind-set. At first Gates was skeptical as well, but he was pleased that the farmer was willing to consult him.

"It was real heartening to get a call from Ralph and to realize this landowner was noticing things in a very sophisticated way," recalled Gates. And when the watershed expert went to the farm and saw what Lentz was so excited about, he realized something important was taking place. "Anybody with two eyeballs could see what was going on. The land was telling us a story."

In fact, what Gates saw at the Lentz farm fit with observations he was beginning to gather in other parts of southeast Minnesota at that time: sometimes controlled grazing of a stream bank helped, not hindered, its stability. It's based on the idea that cattle hooves can be used to create a disturbance in an area for a short period of time—no more than a few days. The ground may look like it was hit by a mud-filled Mack truck immediately after the cattle leave, but it also creates a nutrient-rich environment for new growth to take place.

All that intense impact can also break down the edges of a sharp stream bank, creating a gentler slope for plants to establish themselves. Using short-term livestock disturbance to rehabilitate an area works nicely with the management intensive rotational grazing system Lentz has been using to produce beef since the late 1980s. Lentz reasons that since the animals only stay in the same paddock for a few days at the most, why not make part of a stream bank in need of disturbance one of those grazed paddocks?

Neither Gates nor Lentz are claiming that grazing is the cure-all for what ails a waterway. In fact, they point out several examples of stream banks in farm country that have been improved considerably with the planting of trees and the exclusion of livestock.

Gates feels strongly that if Lentz were tied into one way of raising cattle—an expensive, high-tech total confinement system that left little room for observing and reacting to those observations, for example—he would be limited in how he could manage other aspects of his operation, including the waterway.

"The important thing here is Ralph observes. Ralph could identify fifty plants on his own farm," said Gates one day while sipping coffee on Lentz's back porch, which sits just a hundred yards or so from the banks of Sugarloaf Creek. "He also understands you don't just look at something for one year and draw your conclusions. He does things in little bits and pieces and takes the time to observe the results, and adjust for them. A system like this is never static. It's constantly changing. The key is to remain flexible enough to react to the changes."

References

Barling, R. D., and I. D. Moore. 1994. "Role of Buffer Strips in Management of Waterway Pollution: A Review." *Environmental Management* 18:543–558.

Chapter 7

Stewards of the Wild

Brian A. DeVore

In the early 1990s, a favorite quip of supporters of industrial agriculture was "Farmers were the first environmentalists." By the end of the decade, that phrase had evolved into "Farmers are active environmentalists, not environmental activists."

That's clever: as the ecological costs of modern farming become ever clearer, such common-sensical proclamations have a neat, clean feel to them. They also give a sharp slap to all those "radical environmentalists" out there who wouldn't know a moldboard plow from a field cultivator. The sentiment is right on: Who better to take good care of the land than the people who live and work on it? Unfortunately, agribusiness firms and pro-corporate farming politicians often use such phrases to fend off attempts by the public to have *any say* in how farming is carried out. A document produced by the International Food Information Council shows that "life science" companies like Monsanto and DuPont see a great opportunity in using the positive image of family farmers to promote genetic engineering and other controversial farming technologies (IFIC 1992). A close reading of the paper—it was written for agribusiness insiders who were trying to sway public opinion—shows the survival of real farmers isn't necessarily an important element in these companies' goals. All they need is a mystique that allows for the handy attachment of such mantras as, "farmers are active environmentalists."

This chapter is about a few real farmers who truly are active environmentalists. To these farmers, such an ideal is not ad copy or political rhetoric: it sets the tone for how they live and work on a daily basis. These farmers are proving that food production and environmental sustainability can work hand in hand without the dramatic intervention of chemicals, drugs, and massive amounts of energy. They are also going beyond what the government or perhaps even some environmentalists would expect of them in exercising land stewardship. They're following the land's lead in their management decisions and finding ways of producing food and fiber that are not only economically viable, but also ecologically sound. They are living evidence of what this book maintains: farming as if nature mattered is a viable alternative to industrial agriculture.

A Different Green Revolution

The grass-covered hump in Mike Natvig's pasture is easy to overlook. In fact, only when you are almost on top of it does it become clear what it is: a low-slung dam that traps just enough water in the wet pasture to create a quarter-acre wetland. On a hot July afternoon, the northeast Iowa farmer walked into the shade of a burr oak growing next to the marsh and watched as a startled duck winged its way from water to sky too quickly to be identified. Meanwhile, dragonflies helicoptered from plant to plant, a bullfrog conducted a test of the amphibian broadcast system, and a kingfisher rattled from the branch of another oak on the far side. With all the noise and activity, one wonders how such a biological hot spot could *not* be noticed, even from a distance. This little oasis of water and plants won't be winning any corn yield contests this fall, but it is definitely there for a purpose.

After checking on some chest-high big bluestem growing near the wetland, Mike made his way through the richly diverse pasture. In 1995, he used a no-till planter and the hooves of dozens of beef cattle to seed the 10-acre pasture to native prairie—it is now home to some forty-five species of plants. Natvig crossed a fence onto a neighboring farm that he rents and headed toward a small, spring-fed creek.

He settled his lanky frame down next to the creek at a spot where it wound behind a brilliantly white farmhouse and barn. The creek ran fast and pebbles could be seen on its silt-free bottom, despite the fact that cattle were grazing on its banks just a few days prior. Minnows darted at tiny tufts of shredded grass Natvig tossed at them. The creek flowed through a living-room-sized fen just a few yards downstream. A unique type of marsh that forms peat, fens provide a home for a diverse community of plants, many rare and endangered. In fact, 12 percent of Iowa's endangered and threatened plant species live in fens. Once a farmer knows that and grows to appreciate the plants there, like brilliant blue fringed gentians or the fragrant mountain mint and wild spirea, it doesn't take much effort to protect them.

Seeing such a prime example of a farm operating in tune with nature is even more amazing when one considers the dense piece of modern agricultural history that looms over the area. Norman Borlaug was born in that brilliantly white farmhouse. From those humble beginnings, he went on to win the 1970 Nobel prize for his work in developing high-yielding varieties of grains for the Third World. Natvig's father attended a one-room school with Borlaug, and Mike himself has met the prominent scientist. In recent years, supporters of genetic engineering and other aspects of input-intensive agriculture point to Borlaug's work as a prime example of how science and technology can help feed the world. Critics say Borlaug's "Green Revolution" introduced an unsustainable system of food production to countries all over the world, a system that can't succeed without ever-increasing amounts of chemicals and other expensive and ecologically harmful inputs (Easterbrook 1997).

Whatever Borlaug's place in history, it is clear that the kind of farming Natvig is doing on his family farm is a radical departure from feeding-the-world, high-yield agriculture. And it is real farming—this isn't some sort of untouched wildlife refuge and prairie preserve. He makes a living on the land he farms with his parents, Godfrey and Theodora. He produces hogs, corn, soybeans, and beef cattle on some 400 acres. For his crops, he uses diverse rotations involving small grains and forage plants. He raises organic hogs in a low-cost pasture-farrowing system. Since 1988, he has used management intensive rotational grazing to produce cattle on his pastures, including the one planted to native prairie. This system, which consists of moving livestock in concert with the rate of the growth of the grass, has proven to be an effective method for making soil- and water-holding perennial plant systems pay on a farm. It also allows nutrients in the form of manure to be spread evenly across the landscape at a rate the plant system can make use of (Gerrish and Moore 1995).

Not all of his strategies to manage the land in an environmentally sound manner are as "flashy" as the pasture prairie and its wetland. For example, one spring he planted oats around the first of April. In May, he disced the foot-high oat plants into the soil as he prepared it for planting organic soybeans. Soon heavy rains came—six and a half inches in one week. Erosion was rampant in the area, even though the land in Natvig's neighborhood is relatively flat. But his field of disced up oats held onto its soil.

"If that had been a conventional field without a cover crop, there would have been major washing," said the farmer.

Monitoring is key to Natvig's operation. For example, when he started grazing cattle on the farm he was renting in 1997, he wanted to know what impact the animals would have on the creek that winds its way through the land. The ante was raised when he discovered the fen was also part of the creek.

Using the Land Stewardship Project's Monitoring Tool Box as a guide (see chapter 1), Natvig sampled the stream for macroinvertebrates (small bugs that are key links in an aquatic food chain). At that point, the stream banks had not been grazed for at least ten years. So it was not surprising that more than 80 percent of the insect larvae he identified there were species only found in streams with high-quality habitat. Since then, he has grazed the stream banks twice a year for two days at a crack.

The result? The grazing appears to be having no negative impact on the quality of the stream, and in some ways may be improving it. Natvig knows this because he goes back each year and takes new macroinvertebrate samples with a small net. He's marked his netting spots with fence posts to make sure he gets samples that can be compared from year to year. Recent samples show at least 82 percent of the larvae found in the stream like high water quality. He has also noted that the banks seem to be less sharp and more gently sloped since they were exposed to periodic grazings.

Good land stewardship has taken a priority in Natvig's management decisions since he got started farming in the early 1980s. Burr oaks are his favorite trees, and he has taken pains to preserve the ones that dot his pastures (on the day he showed me his wetland, Natvig was wearing a T-shirt honoring the hardwood tree). He also enjoys canoeing and camping. Natvig, who was born in the mid-1960s, represents that breed of farmer who simply likes wetlands, trees, wildlife, and native prairie. He didn't have to undo a whole lot of management systems when he decided he wanted to farm more sustainably.

"I just farmed based on what I thought was best for the land," said Natvig as he led me back to his house past a line of hazelnut trees he recently planted. "I didn't base it on any government decisions. When I got started farming, I wanted to feel like I could keep the woods and keep places for wildlife. I may not use conservation tillage, but I do have a conservation farm."

Headwaters Thinking

Many pasture wetlands or well-managed streams can add up to one big ecological boon downstream. No one knows that better than Dan Specht. On a late summer evening, several miles south and east of Natvig, Specht wrapped up hog and cattle chores, hopped in his pickup truck and descended to the Mississippi River, just a few minutes' drive away. He had fishing gear in the back, northeast Iowa soil under his fingernails, and nutrient runoff on his mind. That's not unusual. It's difficult for this farmer to separate his various passions—even if they seem to conflict. People concerned about the future of fishing in the Gulf of Mexico would say that farmers like Specht are a direct threat to that industry (both commercial and recreational). Nutrient runoff from Midwestern farms is creating a biological "dead zone" in the Gulf (see chapter 2).

Unlike many farmers, Specht, who's been farming near the Iowa community of McGregor for almost three decades, is willing to shoulder some of the blame for gasping fish in the Gulf. He believes the key is for farmers like him to keep the amount of water and contaminants that leave their fields to a minimum. That's a challenge on the more than 700 acres of steep land that produces crops and livestock for Specht. In these parts, squirrel hunters joke about hiking to the top of backbone-like ridges and pointing their .22-caliber rifles down at the trees, rather than up, and that's not much of an exaggeration.

Like Natvig, Specht produces beef on his steepest ground using management intensive rotational grazing. This means he doesn't have to raise corn and soybeans on his most erosive acres.

On the rest of the land he farms, Specht uses a sophisticated mix of rotations and cover crops. One method the farmer uses is to plant rye in the fall after harvest. By the time the snow melts the following spring, he has a lush, green ground cover that suppresses weeds.

The result of all this effort? A soil surface protected by green vegetation throughout much of the growing season rather than just a few months in the summer. These plants soak up nitrogen as they grow and create a soil structure that stymies runoff.

Such a system can be labor intensive, but it hasn't hurt Specht's production. He recently won a local yield contest with a stand of organic soybeans.

Livestock plays a major role in managing nutrients on Specht's farm. It's difficult to justify the production of small grains like oats and forages like alfalfa, let alone pasture grasses, if there are not hogs or cattle to add value to these crops.

"The system of agriculture where you've got these livestock operations eating the crops they grow on the farm is way more efficient at recycling those nutrients, especially if you can use forages and small grains as part of your rotation," said Specht, who has done on-farm grazing and cover crop research with Iowa State University and Practical Farmers of Iowa. "You're going to be keeping your nutrients where they belong."

The farmer's latest project is a low-cost "hoop house" for raising hogs. This allows him to use bedding from corn stalks and straw from small grains to capture nutrients in the form of manure.

"I'm always working on my nutrient cycle."

Why this desire to zealously control nutrient movement? Part of Specht's concern about what sneaks off his fields is based on the fact that area well water is heavily contaminated with nitrogen, posing a public health threat, particularly to babies. Much of the blame for that contamination can be placed on a Swiss cheese–like limestone geological system called "karst," which underlies northeast Iowa's topsoil. It allows water, and anything that's along for the ride, to easily flow through. That's been shown clearly in Specht's neighborhood by the Big Spring Demonstration Project, one of the nation's longest-running and most detailed studies of the relationships between agriculture and water quality. The research project, which is being coordinated by the Iowa Department of Natural Resources, has found that nitrogen and other agricultural contaminants move quickly through karst geological features into the groundwater. A key component of the research project is the encouragement of farming practices that rely less on heavy tillage and intensive chemical applications (Miller and Brown 1998).

Specht has also had the opportunity to get a big-picture, downstream view of the problem. In the late 1990s, he visited the Gulf and met with commercial fishermen and women whose livelihoods are threatened by excess nutrients from Midwest fields destroying fish habitat.

"It's really fragile. It's vast, but it's fragile," he says of the area where the Mississippi River meets the Gulf.

After making the drive from his farm to the river bottom, Specht pulled into a boat landing below McGregor and met up with frequent fishing partner and fellow farmer Jeff Klinge. While Klinge guided a small aluminum boat out through the backwaters, the two farmers pointed out the natural beauty of the area and talked passionately about fishing. A bald eagle coasted overhead while a great blue heron stood on a point as still as a lawn ornament. Tent caterpillar webs drooped from trees along the water's edge, just a few yards from where a Burlington Northern freight train rattled the Wisconsin side of the bank. Massive barges plied their way up and down the main channel as the farmers began trolling for walleye. This area is vast and fragile as well.

The Lake Effect

Northeast South Dakota farmer Dennis Fagerland likes to talk fishing too. He remembers some of the old timers in his neighborhood telling tales about the slew of fish they used to pull out of Albert's Lake, which adjoins the land on which Fagerland and his wife Jean raise crops and livestock. But the fishing hasn't been that good for quite some time, and when the lake water was tested in the late 1970s, the results showed abnormally high levels of nitrate.

"They said no fish could survive that," recalled Jean.

So the Fagerlands set out to discover the source of the excess nutrients. To their chagrin, they soon learned the contamination was manure runoff from a dairy barn and a beef cattle feedlot—both within a mile of the lake, and both owned and operated by the Fagerlands. This bothered the couple, who have been farming since 1974. Dennis grew up in the area hunting and fishing, and they wanted their five children to enjoy such outdoor activities as well.

Fortunately, the Fagerlands' crop and livestock operation is smack-dab in the middle of a world-class waterfowl nesting area called the "prairie pothole region." Fortunate because that meant there was significant cost-share money (up to 90 percent of the cost of a water-retaining structure was covered in some instances) available from Ducks Unlimited and the U.S. Fish and Wildlife Service to establish small check dams that would catch rain water long enough to provide nesting habitat for ducks and geese.

The structures are little more than culverts with iron gates that can be established in low areas where water tends to flow. In return for the cost-share assistance, the Fagerlands have agreed to keep the gates closed during spring nesting each year.

The seven catch basins (the last one was put in during the early 1990s) serve as areas where water can percolate through heavy, permanent vegetation, filtering out excess nutrients and other contaminants. That vegetation slows and filters the runoff, even when the gates are open.

"That water goes through a lot of cattails before it gets to the lake," said Dennis. "It's pretty well filtered by the time it gets down to the lake. It makes a big difference."

Indeed, recent sonar readings of the lake show fish populations are up. And freshwater shrimp and crayfish are now found clinging to the sides of cage minnow traps set in the lake's shallows. That's a good sign: such bottom-feeding critters feel the brunt of water contamination intimately. Their presence—or absence—can tell a lot about the health of a water system.

"Those crayfish and shrimp—you'd never have seen them before," Dennis told me, obviously impressed at the lake's ability to recover from a near-death experience.

In addition, the flood-control structures slow the water and give the land a chance to absorb it during heavy rains. This is important in that part of the country: The Fagerlands farm on top of the Coteau Hills, a series of ridges that drop 600 feet to the South Dakota lowlands. When the results of a cloud burst starts rolling down that kind of elevation drop, it can cause major flood damage, washing out vegetation and digging deep rills in the side of the ridge (that's why the Fagerlands have been no-tilling their corn and soybeans since the mid-1990s).

The waterfowl are showing their appreciation for these temporary but top-notch wetlands by nesting there in droves. Once while we were talking, Fagerland started to name the species of ducks he's seen in the area in recent years but then gave up in exasperation: "It'd be easier to name the species I *haven't* seen."

The Future

One of the reasons for the lake's recovery is a significant change in the way the Fagerlands farm. In the mid-1990s, they stopped raising beef cattle in the confined space of the feedlot. They haven't used the dairy barn and its outside lot since 1981. In fact, cattle are still being produced on the slopes above the lake, but now they are out on carefully managed pastures. The family is in the midst of converting some 800 acres of land into grazing paddocks for their cow-calf herd. The land was at one time all set aside under the Conservation Reserve Program (CRP). Under the CRP program, the government pays a farmer to basically leave it idle for ten years. This program has been credited with massive improvements in water quality and wildlife habitat in farm country (Feather et al. 1999). But the Fagerlands believe they can get such environmental benefits while actively utilizing the land.

The keystone of their grazing system is a series of "natural" water sources for the cattle. These are in the form of wetlands—each one-half to three acres in size —the Fagerlands have established at the end of each paddock by blocking water flow ("Just add water and you have cattails," quipped Dennis). This cuts down on the amount of walking the cattle need to do to graze, making their feed intake more efficient while reducing the occurrence of compacted "cow paths" and tracked-up watering areas that lead to erosion.

The Fagerlands are confident the wetlands make it possible for cattle and waterfowl to be good neighbors. A myriad of ducks and geese have already found the established marshes and Jean says their experience shows cattle will graze around the waterfowl nests. In fact, the Fagerlands have found that species like Canada geese and pintails prefer the shorter vegetation produced by grazing.

Dennis says that without the "natural" watering holes provided by the wetlands, they probably could not have afforded to set up the grazing paddocks (rotational grazers often have to invest in piping systems to prevent cattle from walking long distances to drink). As a result, the 800 acres would become just one big pasture, which means some parts would have been overgrazed, and others undergrazed. This can be hard on the land and the cattle, while resulting in severe weed problems. Such an inefficient use of grazing areas can have big impacts on the bottom line (Gerrish and Moore 1995).

And this brings up the economics of good land stewardship. Dennis and Jean consider the presence of waterfowl a major bonus to all this. One hundred and twenty acres of marsh habitat wraps itself around the Fagerland house, coming within 50 yards at some points. They both talk about going to sleep at night (and waking up) with the sounds of waterfowl gabbling in their ears.

"I love it," said Dennis one May afternoon as he took a break from field work. "It's really fun around 4:30 or 5 A.M., when everybody wakes up in the marsh. That's quite a sound."

But aesthetics don't pay the mortgage or put food on the table. That's why what farmers like the Fagerlands have managed to pull off is so exciting. The agreement the South Dakota family signed with the U.S. Fish and Wildlife Service and Ducks Unlimited required them to maintain the catch basin gates for ten years. After that, the government's goal of cleaning up water or a conservation group's desire for wildlife habitat may pale in comparison to the farm family's need to turn a profit. The ten-year agreement period on the newest water catchment basin has passed. In early 2001, I asked Dennis, somewhat hesitantly, whether they still left the gates closed during the spring. I shouldn't have been surprised at the answer.

"Yeah, we usually keep them closed until sometime in June. But it gets later every year because you get that water and you hate to see it go," he said somewhat sheepishly.

When the gates are closed, the catch basins will flood anywhere from half an acre (sometimes half an acre is all you need for prime waterfowl nesting) to 23 acres of land. Ten or twenty acres of land under water can be a major sacrifice on a farm. But when the gates are opened, the water leaves behind a rich stand of slough grass. "It's almost like irrigated hay," quipped Dennis. That produces an economic benefit in the form of feed for their one-hundred-head brood cow herd.

A few years ago, the Fagerlands took a close look at one 20-acre spot on their farm that was periodically flooded and found that it produced 8 tons of dry matter per acre — that's triple the average per-acre hay production in the area. That was a pleasant surprise to the Fagerlands, especially after they sat down and penciled out that during ten years of planting those 20 acres to row crops like corn, they had produced only two or three successful harvests.

"Those three years we got crops it was beautiful," recalled Dennis. "But it didn't average out. Then I was able to get a triple hay crop every year on those 20 acres."

Side Benefits

Such side benefits to stewardship may be accidental, but they are also key if such practices are to take hold and spread in farm country. It's obvious that farmers like Mike Natvig, Dan Specht, and the Fagerlands will take that extra step to accommodate nature on their operations. But none of them is independently wealthy, and thus sooner or later financial considerations are bound to come into play in their decision making. Society has learned how to help out with these decisions in some ways—cost-share monies for the wetlands the Fagerlands and Natvig created, for example. In addition, both Natvig and Specht receive price premiums for the crops and livestock they produce because they are certified organic. But many other barriers still exist—both natural and human-based. When Specht converts cropland to grass, government subsidy programs punish him financially for not raising as much corn or soybeans. In order for him to raise chemical-free soybeans, Natvig must plant later in the season to avoid weed problems, making the legumes more vulnerable to frost in the fall. And, let's face it, what if those ducks and geese that nest on the Fagerland farm were instead migrating songbirds? Would sporting groups be as interested in paying for habitat for them?

"Years down the road, I hope we can look back and say it was the right thing to do," said Dennis Fagerland after describing all that his family had done to improve the ecological health of the land. "I think it was."

That's a positive, uplifting statement, but it lacks the hard-as-a-rock confidence of a 1,000-acre corn farmer who's convinced he's feeding the world. The ducks, geese, and even the fish have no such misgivings. As their populations rebound in the Fagerlands' neighborhood, it's become clear that the right thing was done ecologically. Now society needs to figure out how to recognize, support, and reward farmers who live such environmental success stories.

References

Easterbrook, G. 1997. "Forgotten Benefactor of Humanity." *The Atlantic Monthly* 279, no. 1 (January), 75–82.

Feather, P., D. Hellerstein, and L. Hansen. 1999. *Economic Valuation of Environmental Benefits and the Targeting of Conservation Programs: The Case of the CRP.* Report No. 778. April. U.S. Department of Agriculture, Economic Research Service, Washington, D.C.

Gerrish, J. R., and K. C. Moore. 1995. *Economics of Grazing Systems Versus Row Crop Enterprises.* University of Missouri Forage Systems Research Center, Linneus, Mo.

International Food Information Council (IFIC). 1992. *How Americans Relate to Genetically Engineered Foods: A Discussion among Archetype Study Sponsors.* September 14. International Food Information Council, Washington, D.C.

Miller, G. A., and S. S. Brown. 1998. "Big Spring: Farming from the Ground 'Water' Up: Evolution of a Water Quality Project." In *Proceedings of the Sixth Annual Nonpoint-Source Monitoring Workshop, Field Trip* 2. September 23. Iowa Department of Natural Resources Geological Survey Bureau, Des Moines.

Chapter 8

Why Do They Do It?

Brian A. DeVore

When John Lubke fell ill after spraying a cornfield with a chemical that kills rootworms, he thought it was the flu. But then the northeast Iowa farmer came down with the same symptoms the next growing season, again after using the rootworm pesticide. He and his wife Joan started to take note of some other telltale signs that this tool was not good for them or the land. For example, there was the time John found the kinked, lifeless bodies of earthworms all over a cornfield after a spraying of the pesticide (it had rained right after the application). Earthworms are important to the Lubkes, who see these invertebrates as reliable indicators that their soil is in good shape. When I first visited the couple's farm in the summer of 2000, they hadn't used the rootworm pesticide in more than a decade.

Farm country is full of stories like this: dramatic events serving as wake-up calls so loud that they cut through the clamor of industrial agriculture. Strange, scary illnesses associated with agrichemicals are common instigators of change. So are hacking coughs that result from working in livestock confinement buildings, or well-water tests that show nitrate levels so high they endanger the lives of babies. Cattle dying from contaminated feed also gets a person's attention. Farmers have also told me of how they were moved to action by seeing a whole sidehill of rich soil washed away, or mid-January snowbanks taking on a gray pallor because of wayward silt. If a farmer likes to hunt or fish, that opens a whole new field of opportunity for seeing the downsides of industrialized food production: lakes and ponds so full of silt and nutrients that only the lowly carp can survive, fields so devoid of even minimal cover that a relatively minor winter storm wipes out a generation of pheasants and rabbits.

Judging by the number of these traumatic events that occur, one would think that major, systematic change would have already taken place on the majority of our nation's farms, producing a truly sustainable agriculture. In fact, an extensive, eight-state study of agricultural systems that took place in the late 1980s and early 1990s found that sustainable farmers were much more likely to be concerned about such things as water quality and the health effects of agrichemicals (NAF 1994). Much of that concern can be traced to the kind of personal experience John Lubke had. After all, it's one thing when an environmentalist in Washington, D.C., says our food production system has major ecological flaws and that changes are needed to "save the world." It's quite another when farmers have intimate run-ins with such flaws and realize changes are needed just to save themselves and their local community.

But the fact is plenty of farmers get sick from chemicals and do nothing to significantly alter a cropping system that is reliant on those chemicals. Even if they do, say, stop using a particular chemical, that doesn't mean the farm is any more environmentally sustainable overall than it was before. That toxic tool may be replaced with something else that damages the ecosystem in a different way. Responses to a nitrates-in-water scare can take different forms, many of which are very narrowly focused and do not require dramatic changes in a farming operation. Soil erosion caused by extensive row-cropping can be kept in check with massive terracing projects. The nauseating odors produced by livestock confinement operations can be masked with powerful chemicals, special lagoon covers, or a management system that accepts high employee turnover.

Such steps win "Conservation Farmer of the Year" awards and attract cost-share funds and outright financial windfalls from the government, but they don't make the ecological connections required for true change. Dead earthworms and contaminated water are more than isolated problems that can be solved with individual solutions. They are signs of big-picture problems that touch all aspects of a farming operation.

But there are a few shining exceptions. There are farmers who have gone a step beyond and made such connections, using them as a launching pad for bringing about deep, systematic changes on their operations. Why do some farmers see a bad case of the pesticide flu or an eroded field as a call to action that goes beyond utilizing the latest extension-service-prescribed best management practice? There's no ready-made answer to that. It's usually a maddeningly complicated amalgamation of chance and circumstance. But that doesn't mean widespread changes in farming are impossible to direct and support. It just means we have to put in place a farm policy infrastructure that can take advantage of such serendipity (see chapter 18). This book is not proposing any "tweak" of farming methods or government conservation programs but a system-wide conversion based on the ecological understanding that everything is connected to everything else. That kind of thinking will help restore nature on farms. Sustainable agriculture is not simple and neither are the farm-based factors that bring it about.

History Lesson

Perhaps the first lesson to heed is that no one event will truly bring about revolutionary change, no matter how dramatic. We always like to think we can trace back a major life change to one singular moment, one watershed event that sent us in a completely different direction. One day the farm was a conventional operation, the next it was a candidate for a *Sierra* magazine cover story. But it's hardly ever that simple.

Even a short discussion with the Lubkes shows that their stewardship ethic has many sources and branches, many of which predate John's illness or the toxic bath all those earthworms were exposed to. It was evident from the time I entered their long driveway that I was on an operation that paid close attention to stewardship. Running alongside the roadway was a strip of hay that was a few hundred feet wide. Parallel with that was a swatch of corn and alongside that was soybeans. A big sidehill of pasture rose up as a backdrop to these layered, contoured fields, dominating the landscape, putting a capstone on this nice piece of diversity.

I could see this flow of fields out of their kitchen door as I sat at the table and chatted with the couple while they took a break from harvesting oats (actually the break wasn't voluntary; their combine broke down, forcing a delay). The couple farms about 550 acres and has a 100-head beef cow brood herd. All of the land the Lubkes farm is steep enough to be considered highly erodible.

When asked whether water quality is an issue in the area, John answered quickly: "It is to me."

The farmer, who is in his early sixties, explained that early in his farming career he saw a hillside of land get plowed up and planted to corn. The topsoil was severely depleted in just a few years. "I saw what could happen in three or four years. That lesson stuck with me."

It stuck with him even as the government made it increasingly attractive to plant corn and nothing else on those steep hillsides. But being able to see these fields from the kitchen table also means dealing with the not-so-positive messages it sends at times. John and Joan explained how much it pained them that a recent gully washer gouged out a few yards of raw erosion in those soybeans (the erosion would have been even worse if there wasn't the pasture above the soybean field). Indeed, a bit of the erosion could be seen from the kitchen table, a rude reminder that the land doesn't always cooperate with even the most well-intentioned production practices.

"It made me sick," said John, "and we have to look at it everyday."

The Lubkes take pains to use diverse rotations to conserve soil on this steep land and have found ways to reduce chemical use on all aspects of their operation. One way they have done this is by mixing soybean oil emulsifiers with their weed spray. This reduces runoff because the biodegradable oil keeps the chemical right on the plant it is applied to. And because more weed killer stays in place, less is needed to do the job. In fact, the Lubkes have found that by using this method they can cut herbicide rates in half and still control weeds.

Now, if reducing chemical use on the farm was their only goal, they could utilize such isolated best management practices and call it good. But they have a well-rounded stewardship ethic and approach their farming as a way of improving all aspects of the ecosystem.

"We have lots of worms and lots of ladybugs, and maybe we're just trying to make ourselves feel good, but we know from that we aren't harming nature," Joan told me. "People say, 'Well, *we* have earthworms.' But do you see *baby* earthworms?"

Such an attitude makes it easier for farmers like the Lubkes to make a connection between dangerous chemicals, dead earthworms, and eroded soil: all three of these problems might be dealt with separately on other farms. But here, they are part of a bigger whole that involves building the farm's innate fertility and tilth with diverse rotations. Decades may have separated the time John witnessed extreme erosion on a neighbor's farm and the day he became aware of how toxic pesticides can be, but to him it was part of one long continuum.

Deflecting Peer Pressure

The second key for farmers who go beyond simple conservation is walking that fine line between ignoring the neighbors, extension educators, and other sustainable ag naysayers and creating a support network that can help germinate and nurture change.

Dick and Sharon Thompson, whose central Iowa crop and livestock operation is legendary in sustainable agriculture circles, still can't get some of their neighbors to attend field days. Peer pressure is great in all lines of work. But in farming, where the results of a family's work is on display for all to see, the pressure can deflate the healthiest ego. Many farmers, when experimenting with a new production method, hide the test plot in the middle of a field, or at least in a place that's far away from the road. Chemical-free weed control in particular can be tricky with a field full of weeds defiantly waving in the wind. Keeping track of the trials and tribulations of the local weed-hugging farmer can become a spectator sport of sorts. Jaime DeRosier, who farms on the edge of the Red River Valley in northwest Minnesota, uses a complex system of rotations, cover crops, and fallow ground to produce some 1,500 acres of chemical-free crops. His rotation system is highly respected within the sustainable agriculture community. In fact, so many people have approached him for advice that the thirty-something farmer has written a booklet on his system. But one September day as we hiked across a three-week-old stand of hairy vetch and winter rye (such a seeding suppresses weeds, creates fertility, and protects the soil from erosion) growing in one of his fields, DeRosier talked about how the three or more years it can take to take a field from chemical-dependent to biologically independent can be difficult—agronomically and psychologically.

"Transitioning land out of chemicals, it seems to fall on its face the first year because it was getting all those chemicals, and it's dependent upon them. People are always watching me pretty closely because they know I'll be transitioning once in awhile and will fall on my face."

Some farmers deal with these assaults on self-esteem by avoiding the coffee shop altogether. Still others meet their critics head-on with a self-deprecating sense of humor. When they first started raising soybeans without chemicals, southern Minnesota brothers Don, George, and Ray Yokiel used to tell neighbors, "We're getting 60 bushels to the acre—40 of beans and 20 of weeds."

Sometimes the most powerful pressure comes uncomfortably from close at hand.

Since 1969, Dave and Florence Minar have been farming near New Prague, a community about 40 minutes south of Minnesota's Twin Cities. The land they farm was where Dave grew up and his family has a lot of deep roots in the community. Like John Lubke, Dave had a run-in with a farm chemical one day in 1973. After mixing a tank of herbicide, he got a "real confused, nervous feeling, and I had to lay down." Dave said it may not have been the chemicals that caused the illness, but it reinforced the concerns he and Florence had about them just the same. They went almost cold turkey on chemicals. It wasn't an easy decision, recalled Dave—especially considering the family's history in the community.

"My father was the first in my community to apply herbicides, and we were the first to stop using them."

Dropping chemicals so suddenly produced some pretty rough-looking fields in those early years. Dave's mother would say: "Look as those weeds in the corn, what are the neighbors going to say?"

Minar is almost apologetic for his willingness to ignore peer pressure: "It's part of my personality, I guess. I would say, 'I don't care what the neighbors say.'"

Support Networks

Let's not get the wrong idea here. Farmers who successfully make major changes in how their farm relates to the ecosystem do not inhabit some sort of agronomic fortress of solitude. In fact, networking with other farmers is critical to the success of transitioning into sustainable agriculture. Studies have found that farmers who are part of some sort of sustainable producers' organization or informal network were more likely to be successful economically and ecologically (NAF 1994). Those same studies have shown that the resource farmers said they most lacked when seeking to adopt more sustainable methods was information. That's not surprising, considering how sustainable practices tend to be more management intensive than conventional systems.

Dave and Florence Minar have long been involved with the Sustainable Farming Association of Minnesota and the Land Stewardship Project. DeRosier is also involved with a local Sustainable Farming Association chapter. The Lubkes do on-farm experiments in conjunction with the Practical Farmers of Iowa and the Michael Fields Agricultural Institute.

One of the most exciting forms of farmer networking has grown up around management intensive rotational grazing. During the latter part of the 1990s, dozens of grazing clubs starting popping up in Wisconsin, Iowa, and Minnesota. Patterned after similar groups in New Zealand, these groups were in response to the realization that when farmers start grazing they have entered a world where animal behavior, perennial plant growth, and soil biological activity interact in sometimes wonderful, oftentimes baffling, ways. Decisions based on the interactions among tractors, herbicides, and hybrid seed don't apply anymore.

"There's no comparison," said southeast Minnesota farmer Joe Austin one June day in 1999. As he said this, several members of his grazing club and I were trekking through the hilly pastures on which he and his wife Bonnie produce cattle and sheep. Many of their pastures had been converted from row crops just a few years before. "It's a different thought process. Once you put the corn in, everything is cut and dry until you market it. With grazing, every day is different. No two years are the same. You think you have it whipped and then something else hits you."

These grazing groups usually meet once a month on a member-farmer's operation. Some even hire a coordinator who has some sort of background in farm business management and grazing. They often have confidentiality agreements so that farmers feel comfortable discussing such ticklish subjects as finances.

"My grazing group has gotten to be a very important part of my life," said Dave Minar.

As sustainable agriculture coordinator for Iowa State University Extension, Jerry DeWitt has studied the dynamics of such groups. He believes such teams of farmers can help sustainable agriculture management techniques spread through rural communities at the grassroots level. That's different than the traditional model of spawning "innovations" in farm country. Under that model, land grant extension services pick out a respected farmer in a certain county or region and have that farmer host an on-farm experiment. After the experiment is established, a field day is held, exposing the innovation to dozens of farmers at once. In theory, farmers then go home and attempt to mimic what they saw. But there are a couple of problems with that model, according to DeWitt. First, it doesn't take into account the nuanced, but often critical, differences between farms—even ones that may be right across the road from each other. It also doesn't give enough credit to the innate ability of farmers to develop their own innovations.

"This kind of rote dissemination of information has became common during the past several decades," said DeWitt. "But now we're getting beyond the simple model of 'I'm going to be just like that person.' I think people want their information from multiple sources. It's a different learning model than the 'an expert tells you what to do' model."

Consumer Support

Many innovative sustainable farmers have discovered the value of networking with a group of people they normally wouldn't run into at an agricultural field day: consumers. This has come about by accident, mostly, as farmers using sustainable methods have become frustrated with the prices the conventional food system pays. Some have been able to capture more profit through organic premiums. But others have gone straight to the shoppers, marketing meat, dairy products, and produce through farmers' markets, farm pick-ups, Community Supported Agriculture, and other forms of marketing that rely on personal relationships between farmers and consumers. (See chapter 17 for a broader discussion on marketing.) This form of marketing starts out as a way to capture more of the consumer dollar. But for farmers who do it well, it is also a network that is just as key as any grazing club. It provides moral support for the type of farming they are doing as well as feedback directly from eaters on what works and what doesn't. The consumers, for their part, learn about the challenges family farmers face in raising safe, ecologically sustainable food. Such information can come in handy when they shop as well as when they vote or otherwise play a part in decisions that affect agriculture's role in society.

In 1993, the Minars converted to grass-based livestock production and started making a serious foray into direct marketing of sustainably raised beef, pork, chickens, and turkeys. In 2001, they broke ground on a milk-processing facility right on their farm. They are hoping to use that facility to direct market their milk in the form of yogurt, cheese, and other products.

They now sell pork, beef, chickens, and turkeys to more than two hundred customers, who pay a premium price. Milk sales still make up most of their income, but by 2001 direct-to-consumer meat sales were about a quarter of their gross profits. The Minars are well known for their devotion to producing food in an environmentally sustainable manner. In 2000, they joined with area conservation agencies and nonprofits in a project to reclaim a highly eroded portion of Sand Creek, which winds its way through their farm. The project is attempting to prove that low-cost methods using excavating equipment, logs, cattle, and grass seeding can improve a stream bank cheaply and effectively. The project has some major implications. Sand Creek is one of the main sources of siltation in the heavily polluted Minnesota River (Mulla and Mallawatantri 1997). In addition, it's believed by an increasing number of ecologists and conservation technicians that low-cost methods of ecosystem reclamation are what are needed in farm country—methods that utilize the farm's own resources and tools.

I've been to the Minars' many times when their customers pick up meat from their walk-in freezer or attend an open house that features not only the farmers, but also the people who do the meat processing for them. The Minars consciously do not deliver any of their products. They want their customers to see their farm and meet their family, which includes five children. These customers, many of whom are recent immigrants from the Twin Cities, show their support for this kind of agriculture with their pocketbook. But they also provide verbal pats on the back and end up telling others about this wonderful source of family-farm meat.

Given their environmentally friendly reputation, I asked Dave if that pays off in their direct marketing. Even though they tout their eco-methods in brochures and sales material, Dave said in the end the thing most consumers are interested in is whether the meat is chemical and antibiotic-free (it is).

That the people Dave and Florence Minar sell sustainably raised meat to are first and foremost focused on their own health is understandable. After all, that's what has prompted many a hard look by farm families themselves over the years. But just as a case of the pesticide flu can be one of the factors that leads a farmer to look at the bigger agro-ecological picture, consumer concerns about food safety offer a teachable moment, a door into a wider understanding about the importance of supporting sustainable farming systems. For example, the Minars are able to raise drug-free beef because the cattle are out on well-managed pastures where health problems related to confinement are not present (Goldberg et al. 1992). Those pastures keep contaminants out of the air and water, and recycle nutrients efficiently without the use of mega-manure lagoons. And the low capital costs of such a system allow a family like the Minars to live on the land and make a living from it.

Gaining the moral support of residents in the area is becoming more important to the Minars literally by the day. According to the 2000 U.S. Census, their home county, Scott, is the fastest growing in the state. In fact, local government officials project that by 2010 Scott County will have 109,000 people living there—up from less than 60,000 in 1990 (Bureau of the Census 2000). Subdivisions are sprouting all around the Minars' pastures. This concerns the family, but they aren't about to move. In his typical laid-back way, Dave told me they see such growth as "bringing our customers closer to us." And if those customers can associate agriculture with a direct source of food, as well as beauty (the Minars often don't have to provide directions to their place to even newcomers because their verdant pastures are so well known) and clean water, so much the better. But it's not just about safe food, clean water, or scenic vistas. The Minar farm fits into a larger ecosystem that blends all of these factors and more. This ecosystem is many things to many people, but as far as Dave and Florence are concerned, at the core of it all is that it is their family's home.

"We intend to stay here, and part of it is having the animals out and not contributing to the smell and being a good neighbor," said Dave. "And if that means providing meat directly to consumers then that's part of it too."

All the Pieces

When a farm is set up to put individual problems in the context of an interconnected whole, it can be a beautiful thing to watch. One of the best recent examples of this in action is the story Tom Frantzen tells about his family's decision to raise hogs in deep straw bedding. Tom and his wife Irene are very well connected in sustainable agriculture circles. The New Hampton, Iowa, farmers have long been involved with Practical Farmers of Iowa and the Land Stewardship Project. He has attended conferences across the country and visited other farms. Tom makes regular deliveries of pork to a natural foods co-op in the college town of Decorah. During deliveries, he takes the time to discuss sustainable food production with the co-op manager as well as with the customers. The Frantzens' connections extend into the scientific community as well: they've done research in collaboration with Iowa State University and the University of Northern Iowa.

All of these connections have served the Frantzens well over the years as they've made extensive changes to their operation in an effort to make it more sustainable. One of the last bastions of conventional agriculture on their property was their hog operation. For fourteen years, the Frantzens raised hogs in the same way many of their neighbors did: in closed buildings with concrete floors. The floors had special slots in them so that urine and feces could drain down into a pit below. All this liquid manure had to be pumped out and disposed of. Such a system was bad for the stressed-out animals (they fought each other and required lots of antibiotics) and the environment (liquid manure often finds its way into waterways), as well as members of the Frantzen family (who had to work in facilities full of dust and toxic gases). In short, recalled Tom one day while taking a break from chores, this system treated animals as machines, manure as waste, and farmers as barnyard janitors.

But in 1997, the Frantzens junked the trappings of confinement and started raising hogs in deep-straw bedding in open-ended, Quonset hut-like structures called hoop houses. The family was already raising hogs on carefully managed pastures in the summer, but producing pork during harsh Midwestern winters meant the hogs had to be confined—or so the Frantzens thought until they visited Sweden. There they saw pigs being raised under natural conditions using deep straw bedding. The bedding allows manure to break down in a dry system rather than in a liquid, anaerobic environment.

The family could see the advantages to this system from the start. For one thing, it could be set up for about one-fourth to one-third of the cost of a confinement facility. In addition, the pigs were healthier because they were allowed to follow their natural instincts to socialize and nest. Finally, when the manure mixed with the straw, it created a composting "pack" that kept the animals warm. The ecological benefits of these systems are excellent: because they rely on a solid manure system, hoops don't have the inherent environmental risks associated with large liquid waste pits and lagoons. Once a straw/manure pack is scraped out of a hoop house and piled up, further composting reduces the mass to half its former size, according to Tom. It can then serve as biologically rich, organic fertilizer for farm fields. No more dealing with liquid hog manure as a waste. Plus, Tom didn't feel good about the inability of his hogs to move around and be more hog-like.

"Every time I observed the crowded, stressed pigs, I too became stressed—their social brutality was caused by my failure to meet their basic instincts," recalled Tom as he sat on his four-wheeler in the shade. "On a hoop-building tour, I was told that pigs have three desires: they want to run around, build a nest, and chew on something. This behavior is impossible in a building with metal pens and slat floors."

Still, Tom was apprehensive about making such a significant switch from a system that had the agri-science seal of approval. His concerns were put to rest one September day in 1997. The hoops they were building weren't quite finished yet, but he decided to move one group of pigs anyway—that morning he had opened the door to one of his confinement sheds to find blood all over the place. The stress levels in the pigs had built to the breaking point and they were cannibalizing each other. He covered the floor of one of the hoop building with fresh straw and turned 160 pigs loose.

> Boy did those pigs have fun. In the new hoop building, they had twenty-one hundred square feet of room to run, straw to chew, and lots of bedding to nest in. They ran around all day and into the night. The next morning I ran out to check on them and I will never forget what I found. As I walked up to the open door, it was quiet, very quiet. I peeked into the hoop house to see 160 pigs in one massive straw nest, snoring with great content. I laughed until I cried. Their stress was gone, and so was mine.

I like that story because it shows how elements like personal discomfort, a concern for the environment, financial issues, family quality of life, and accessibility to information on an alternative can help create true changes on a farm. The Frantzens didn't just decide to raise hogs differently using a new kind of gadget—they set in motion a whole series of events. Hoop buildings, or any other off-farm products for that matter, aren't environmental silver bullets. Indeed, if one were to concentrate enough hoop houses in one place, environmental problems would surely crop up. Rather, they just happen to dovetail nicely with a system of farming that considers all of an operation's enterprises and resources as part of a larger whole. The compost produced by the hoop house system is a biologically rich fertilizer, which farmers like the Frantzens use on their crops. Those crops are then fed to the animals and cycled back to the land as manure, helping to produce more feed crops. In addition, the straw itself comes from small grains such as oats, a soil-conserving crop that can naturally break up weed and insect pest cycles. In 1995, the owner of a large-scale hog factory confronted me as I stood on a gravel road looking at his multi-million gallon liquid manure lagoon. His operation hauled in corn and soybeans, and shipped out millions of gallons of liquid feces. Earlier in the day, he had been on a farm that was raising hogs in deep-straw bedding. "If we started raising hogs like that, every farmer in the country would have to be raising straw," he yelled with no sense of the positive implications of his argument.

And pasture-farrowing hogs during the summer—along with grazing cattle—helps the Frantzens economically justify having a large portion of their farm in perennial grasses, which add biological diversity to the landscape.

An environmental audit done in 1999 on the Frantzen farm by the University of Northern Iowa found the operation was producing between only 20 and 30 pounds of excess nitrogen per acre annually, a sign that it's extremely efficient at keeping wayward nutrients from becoming pollutants. In comparison, nitrogen loss on liquid manure–based factory operations is measured in the hundreds of pounds per acre.

Data like that just reinforces what those happy pigs had already expressed to Tom on that September morning: this is the right way to produce food. More than three years after he started using the hoops, Tom claims, "I've been on the other side of the fence and I will never go back."

How Do We Promote It?

Do stories like this bring us any closer to figuring out how to make the Lubkes, DeRosiers, Minars, and Frantzens of the world the rule, rather than the exception? As I mentioned at the beginning of this chapter, the ways in which farmers make major sustainable changes in their operations are not easy to put in a how-to manual. Farming, if done right, should be a creative process. But how do you nurture the creative process without smothering it?

Perhaps we just need to create the right environment in which agricultural serendipity can thrive. Current federal commodity policies, coupled with land grant research agendas and "feed the world" food markets only serve to constrict the choices farmers have. Anything that can be done to propagate farmer-to-farmer support networks, large and small, formal and not-so-formal, is critical. We also need to fill the information void through more funding of research that fits the needs of farmers rather than of input suppliers. What if the Frantzens had not already viewed alternative swine production firsthand when they started to have questions about their old system? Would they have simply stopped raising hogs, or opted for a high-tech confinement system?

We also need to redefine what it means to be a "Conservation Farmer of the Year." That means viewing farms not as industrial islands where inputs—including expertise—are shipped in, and waste products are shipped out. Rather, these farms, and the people who own and operate them, are as much a part of the landscape as the most beautiful state park.

We can start making these farms part of the landscape by seeing them through the eyes of someone like northeast Iowa cattle farmer Greg Koether, who recently told me what he feels when he sees wildflowers growing among his oaks, turkey strutting in the pastures, or the precipitation from severe rainstorms staying put on his grazing pastures: "It's a real joy."

References

Goldberg, J. J., E. E. Wildman, J. W. Pankey, J. R. Kunkel, D. B Howard, and B. M. Murphy. 1992. "The Influence of Intensively Managed Rotational Grazing, Traditional Continuous Grazing, and Confinement Housing on Bulk Milk Tank Quality and Udder Health." *Journal of Dairy Science* 75:96–104.

Mulla, D. J., and A. P. Mallawatantri. 1997. Minnesota River Basin Water Quality Overview. FO-7079-E. University of Minnesota Extension Service, St. Paul, Minn.

Northwest Area Foundation (NAF). 1994. *A Better Row to Hoe: The Economic, Environmental, and Social Impact of Sustainable Agriculture*. December. Northwest Area Foundation, St. Paul, Minn.

U.S. Bureau of the Census. 2000. "Basic Facts–Quick Tables: Scott County, Minnesota." *Census 2000 Redistricting Data*, U.S. Bureau of the Census, Washington, D.C.

Part III

Ecosystem Management and Farmlands

In the early 1990s, federal agencies such as the U.S. Forest Service set out to change the way they managed public lands. At the insistence of environmentalists from both without and within, these agencies began to think bigger than their own boundaries and economically valuable natural resources and to look instead at *whole ecosystems*. They were forced to. For one thing, some of the endangered species they were legally bound to protect, such as grizzly bears and spotted owls, did not see these boundaries—only habitat. Furthermore, large-scale processes so important to forest and rangeland ecology, such as fire, required a long-range, large-scale ecosystem perspective. After decades of the multiple-use philosophy, such a change was difficult. It was particularly hard to accept that logging, grazing, and mining were not simply the business of natural resources scientists working for government agencies and industry but were in fact every citizen's business, a legitimate topic of public discourse. The new philosophy, called *ecosystem management*, has been described as "integrating scientific knowledge of ecological relationships within a complex sociopolitical and values framework toward the general goal of protecting native ecosystem integrity over the long term" (Grumbine 1994).

Today, agricultural states are facing a predicament similar to that which the U.S. Forest Service faced ten years ago. Many endangered species have requirements that simply cannot be met without drastic changes on a landscape scale. The Topeka shiner is a small fish that needs clear, gravel-lined prairie streams. It has been extirpated from 80 percent of its formerly broad range due to stream channelization and excessive siltation (much of that loss occurring in the past twenty-five years). The pallid sturgeon, an ancient, primitive fish of large rivers, requires occasional spring flooding in order to spawn. But this is something the Missouri River is no longer allowed to do, since it would interfere with farming in the floodplain and barge transportation. Clearly, neither of these fish can be saved by "fish preserves" but will instead require system-wide changes in management of land and water if they are to recover.

Agricultural states face difficult choices when it comes to water quality. Up until recently, the states didn't attempt to enforce clean water standards for fertilizers, pesticides, and sediments leaving farm fields because these were considered nonpoint source pollution. Instead, the states financed demonstration projects and encouraged "best management practices." Now, the Environmental Protection Agency is forcing states to clean up hundreds of streams and lakes that do not meet minimum clean water standards. Clearly, the permitting process used for point source pollution would be impossible to implement on farms—a public relations and enforcement nightmare. Only an ecosystem approach will really address this problem, but how can this happen when land is managed not by a federal agency under the control of taxpayers but by hundreds of thousands of independent, private property owners all pursuing the American dream?

The chapters in Part III provide a glimpse of what ecosystem management could look like in the context of the intensively farmed, privately owned landscape. In chapter 9, Heather Robertson and Richard Jefferson of English Nature describe the way native biological diversity thrived under certain forms of traditional farm management before World War II. Since then, various government policies, as well as the influence of world markets, have reduced the amount of habitat available. Recognizing that the rural landscape provides many values to society besides food, the British government has created popular incentives for farmers to continue to manage their land in ways that maintain or restore native plants and threatened wildlife habitat. Due in part to a deep tradition of natural history studies in the rural countryside, and partly due to an understanding that farming can not be separated from the natural world, British policies appear well ahead of those of the United States in ecosystem management of farming country.

In chapter 10, Laura Jackson compares the ecosystem properties of agroecosystems in Iowa to the prairie ecosystem that preceded them. The agroecosystem that dominated from the 1860s until the 1950s preserved many characteristics of the prairie ecosystem, characteristics that the corn-soybeans system lacks. A return to long crop rotations, along with significant innovations in tillage and grazing technology, and aggressive efforts to preserve remnant tracts of native habitat, is proposed as a form of agroecological restoration.

Nick Jordan examines in chapter 11 how biological diversity helps to make agricultural ecosystems both more productive (in terms of yield) and also more sustainable. Soil health and pest management are two areas in which ecological partnerships among diverse organisms are needed. Such beneficial ecosystem processes can only be used on farms if they are sought out and actively managed by farmers.

A primary realization leading to ecosystem management is that conservation is not carried out in isolation. Thus, The Nature Conservancy began in the early 1990s to think beyond individual species or preserves to landscapes and ecosystems. As a result, it had to get to know its neighbors. In chapter 12, Judy Soule reviews eight different conservation projects that The Nature Conservancy is carrying out across the United States. In some cases, conservation requires extending incentives for farmers to change their practices; in other cases, farmers may benefit economically from changing their practices to benefit wildlife. In each case, however, good communication between The Nature Conservancy and its neighbors is critical.

In chapter 13, Carol Shennan and Collin Bode describe a project in the Tule Lake National Wildlife Refuge, a unique "dual function" refuge with a federal mandate to promote both waterfowl and farming. Both farming and wildlife have been in trouble lately. High-quality wetland habitat has shrunk, possibly due to the lack of fluctuating water levels and sedimentation. Farmers leasing lands within the refuge have suffered poor prices and high pest populations in fields farmed for years. Refuge staff are experimenting with rotating potato production with seasonal wetlands on both short and long cycles. Such a rotation would partially mimic the fluctuating water regimes that characterized the natural wetlands. Shennan and Bode conclude with an extensive discussion of the barriers to adoption of this system, including society's general reluctance to mix farming with wildlife conservation, and difficult regionwide conflicts over water.

References

Grumbine, R. E. 1994. "What Is Ecosystem Management?" *Conservation Biology* 8:27–38.

Chapter 9

Nature and Farming in Britain

Heather J. Robertson and Richard G. Jefferson

The ramparts of Barbury Hill Fort shimmered in the hot July sun. The small herd of beef cattle watched us intently. It was the dog of course. The cattle had scarcely glanced at an approaching group of heavily laden and overheated hikers who were toiling up the ancient Ridgeway toward the fort. After subjecting us to a few moments of motionless scrutiny, in communal decision the cattle resumed grazing. Our whippy black hound was on a leash and could safely be ignored. The standoff over, we scrambled up the steep grassy ramparts of the fort. From there, high on the chalk crest of the Marlborough Downs, we could see into the blue distance across the wide floodplain of the River Thames. This was the landscape that the Iron Age builders of the fort must have seen from their dazzling white stronghold, freshly dug from the chalk, in the time before the Roman invasion in A.D. 43. Two thousand years later, the fort remains as a double circle of steep embankments in a ring about 100 meters in diameter. The ramparts and the deep ditch between them seemed to us a formidable defense, but the fort was no match for Roman military might and was abandoned by its inhabitants.

In the peaceful present, a delicate tapestry of flowers covers these grassy relicts. The tiny pale pink petals of squinancywort are highlighted by the deeper pink blossoms of wild thyme, the lilac of small scabious, and the sulfur yellow of rock rose. After the fort was deserted by its human occupants, wild plants of chalky soils spread on to the ramparts from the surrounding man-made prairies. These had been created by forest clearance as long ago as the Neolithic period and the Bronze Age. Centuries of grazing by livestock have kept the fort's grasslands free from encroaching trees, as the cattle are still doing today. Chalk grasslands like this are found on the rolling chalk hills and their ancient earthworks all across southern England. They are among our richest habitats for wildlife. Upward of forty species of plants are crammed into every square meter and insects abound. The brilliant Adonis blue butterfly is a visual treat, while grasshoppers and crickets buzz in the background. The soaring song of the skylark is a reminder that birds of open country thrive in this farmed landscape.

The Intertwined History of Farming and Nature

On the flat ground beside the River Thames at Cricklade, upriver from Barbury, is another grassland that exemplifies the long historical connection between farming and nature in Britain and also illustrates the economic and cultural dimensions of this relationship. Grassland has been the predominant vegetation on the Thames floodplain since forest clearance in prehistoric times. Here the Romans made hay from the lush herbage growing on the rich alluvial soils (Lambrick and Robinson 1988). Cricklade itself was an Anglo-Saxon settlement founded in 890. Its inhabitants had legal rights to cut hay from the meadowland around the town. These rights became codified into the Lammas Land system, presided over by a court that appointed a Hayward to supervise management of the hay meadows. Under the system, the land was grazed in common by livestock owned by right holders from Lammas Day (August 12) to Lady Day (February 12), then animals were removed. In the summer the hay crop was harvested in strips by individual right holders. This medieval land management arrangement persists today in North Meadow, which lies on the western edge of the town. The 44-hectare meadow has a rich flora of wild grasses and herbs, including over a million snake's head fritillaries (Payne 1998).

The economic value of such meadows in Britain in historical times is undisputed. Hay was vital winter food for the horses and oxen that provided the power for spring plowing. Meadowland was the most recorded land use in the census known as the Domesday Book, which was compiled in 1086. A monetary valuation of an Essex estate in 1309 put meadow at four times the value per acre compared to pasture and eight times the value of arable land (Rackham 1986). North Meadow is a typical example of an economically important agricultural land use but also demonstrates how the interrelationship of farming and nature entered Britain's cultural consciousness. It is symbolic of the meadows vividly described by Shakespeare in his play, *Henry V*.

> The even mead, that erst brought sweetly forth
> The freckled cowslip, burnet and green clover,
> Wanting the scythe, all uncorrected, rank,
> Conceives by idleness, and nothing teems
> But hateful docks, rough thistles, kecksies, burs,
> Losing both beauty and utility.

Shakespeare realized that the wild meadow plants, highly valued as hay, benefited from farming and that without continued management they would succumb to competition from worthless weeds.

Pastures of chalk uplands and meadows in river valleys were only two elements of a farmed landscape that evolved over centuries in Britain, a process eloquently described by Oliver Rackham in his book *The History of the Countryside* (1986). The essence of this countryside is summed up in one line from the poem "Pied Beauty," by Gerard Manley Hopkins: "Landscape plotted and pieced—fold, fallow and plough." Mixed livestock and arable farming was the predominant system and by happy accident provided abundant niches for wild plants and animals. Newts and toads found living space in the ponds dug as water supply for livestock, hedges planted as stock-proof field boundaries sheltered song birds and dormice.

Farmland habitats often had multiple economic uses apart from forage and food crops. As well as rough grazing, heathlands and moorlands yielded spiny gorse and low, shrubby ling for fuel; bracken fern for animal bedding and soap making; and the self-explained dyer's greenweed. Wild animals thriving in these habitats included smooth snake, nightjar, golden plover, and red deer. The marshland home of swallowtail and large copper butterflies supplied reeds for thatching roofs, and on the coast, fuel for salt making.

Before the advent of modern medicine, the countryside was also the pharmacy. One example was meadowsweet or queen-of-the-meadows. The creamy foam flowers of this plant were a characteristic sight in damp meadows. *Gerard's Herbal* of 1597 recommended a distillation of the plant for treating burning and itching eyes and extolled the scent of meadowsweet because it "makes the heart merrie" (Woodward 1994). The story continues into more recent times because in the late nineteenth century a precursor of aspirin was extracted from Meadowsweet. The drug itself was apparently named after the old Latin name for meadowsweet, *Spirea ulmaria* (Mabey 1996).

The diverse countryside of Britain, derived from a farming economy and supporting a rich wildlife, was celebrated not just in poetry but also in music and art. One supreme example is John Constable's *Haywain,* painted in 1821. Its subject, a horse-drawn hay wagon standing in a pool of sunlight in the luminous landscape of Dedham Vale, is probably the best-known painting in Britain.

The lineaments of the landscape were not totally fixed through time. The proportions of grassland and arable fluctuated along with national fortunes. Grassland replaced arable in many places when the Black Death reduced the human population by a third in the fourteenth century while the arable area expanded again at the time of the Napoleonic Wars. Grassland increased once more when cheap grain began to be imported from North America in the

1870s (Duffey et al. 1974, Rackham 1986). The Industrial Revolution in the mid-eighteenth century and the steeply rising population led to the urbanization of considerable tracts of countryside (Hoskins 1955), but even today over 70 percent of Britain is farmed.

The British countryside has been studied by a long succession of naturalists, originating in the explorations by apothecaries searching for medicinal plants. The earliest recorded excursion of the Society of Apothecaries of the City of London was May 1620, the year the Mayflower set sail for America (Allen 1976). A more recent example of organized recording was the production of the *Atlas of the British Flora* in the 1950s through the combined effort of 1,500 volunteer botanists (Perring and Walters 1962). The public interest in British wildlife continues unabated. The Royal Society for the Protection of Birds has over one million members, while the recently produced *Flora Britannica* (Mabey 1996) was a best seller. The massive amount of information collected by naturalists over the years means that there is abundant evidence for the importance of the farmed landscape for wild plants and animals and for the significance of changes in farming practices in the twentieth century.

The Impact of Farming on Nature after the Second World War

The last soldiers to camp at Barbury Hill Fort were part of the United States Army forces gathered in Britain for the invasion of Europe in 1944. The cataclysm of the war in which they were fighting marked a turning point in the relationship between farming and nature in Britain. Before the war, Britain imported 70 percent of its food. The threat of starvation loomed during the war as more and more ships in Atlantic convoys were sunk by enemy submarines. Many ancient grasslands were plowed up by government order in a desperate attempt to grow more crops, although some farmers defied the orders and risked jail to save their meadows. After the end of the war, the government was determined that Britain should be self-sufficient in food in future. Subsidy money, research, and agricultural advice were all part of the drive to increase production by intensifying farming practices. Land drainage, inorganic fertilizers, chemical herbicides, and pesticides all increased crop and forage yields. Farms became more specialized into all arable or all livestock enterprises (O'Connor and Shrubb 1986). Britain joined the European Community in 1973, and production subsidies under the Common Agricultural Policy further stimulated intensification throughout the remaining decades of the twentieth century, leading to the creation of surpluses across Europe. These were derided by the British press in catchy phrases such as "the butter mountain" and "the wine lake."

The intensification of farming created a tidal wave of destruction that swept away wildlife habitats across the country. This loss is illustrated in Figure 9-1, which shows the 88 percent reduction in chalk grassland on Chanctonbury Hill, West Sussex, between 1947 and 1991. Only small fragments of agriculturally "unimproved" grassland remain from a once continuous area. The gentler slopes were plowed while scrub spread on steep slopes because of a lack of livestock grazing. Across England and Wales 97 percent of Shakespeare's meadows have disappeared, their flora destroyed by the "improvement" process of heavy applications of inorganic fertilizers, which favor only a few competitive species (Fuller 1987). Agricultural improvement was the main factor behind the 40 percent loss of heathland between 1950 and 1984 (Nature Conservancy Council 1984). Farms growing only arable crops had no need for ponds and hedges that hampered the movement of the large modern machinery that had replaced horse and oxen power. Between the end of the war and 1974, 120,000 miles of hedge were removed to facilitate agriculture (Nature Conservancy Council 1984). The wild plants in the arable fields themselves also declined sharply under the combined impact of fertilizers, herbicides, modern seed cleaning techniques, and the change from spring to winter sowing of crops. Most of the wild plants are annuals that germinate in spring and suffer from the greater competition from crops sown in the winter rather than the spring. The bright blue cornflower was recorded from three hundred and seventy-four 10-kilometer squares in Britain before 1970 but only eighty-two squares between 1970 and 1986. After this date, it has only been found in a handful of arable fields (Wigginton 1999).

The widespread destruction of wildlife habitats was not intentional but was a most unhappy accident resulting from changes in farming economics and technology. Nevertheless, there was a growing realization among the British people that something priceless was being lost, culminating in impassioned pleas for the protection of the countryside. One influential example was Marion Shoard's book called *The Theft of the Countryside* (1980). She ended her last chapter with an appeal to the democratic process. She was convinced that the countryside could be saved "if the people will it."

Farming and Nature Today: A Visit to Applesham Farm

On a golden September day after harvest, we toured the farm with Chris Passmore, who is the third generation of his family to farm the land here. Most of the 344 hectares of Applesham Farm lie hidden in a coomb, or dry valley, encircled by the smooth chalk slopes of the South Downs in West Sussex. Chris runs a profitable mixed farm enterprise comprising sheep, beef cattle, and grain production. Arable crops are rotated with three-year "leys" of sown grass and clover. Permanent chalk grassland covers 30 hectares of the steep slopes along the southern and western borders of the farm. Chris has a deep knowledge of the many and varied wild plant and animal inhabitants of his farm, some of which we were lucky enough to see on this perfect September

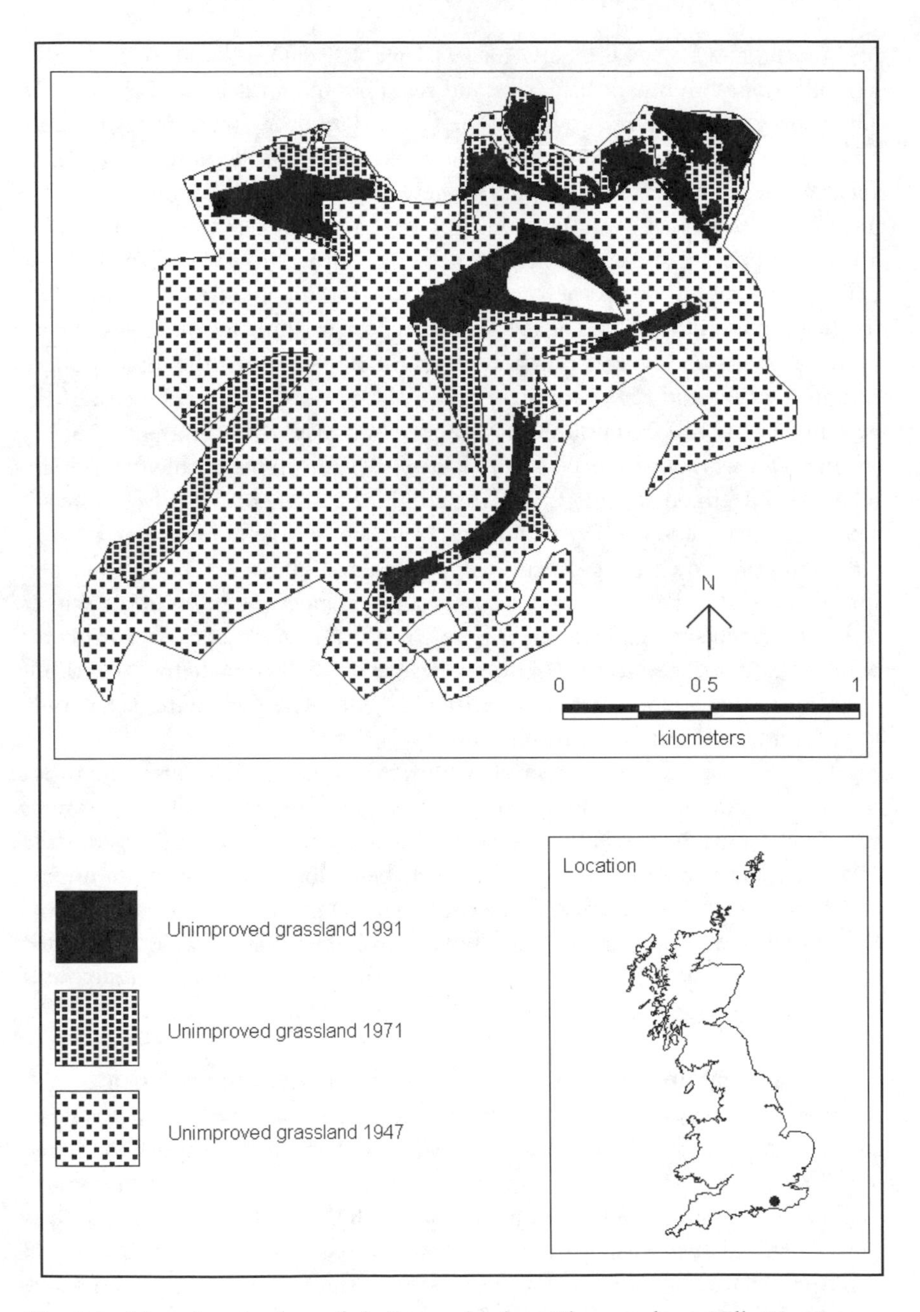

Fig. 9-1. Map showing loss of chalk grassland on Chanctonbury Hill, West Sussex, U.K., between 1947 and 1991.

day. Highlights were the covey of grey partridge, which whizzed away from us as we walked over the fields, the clouded yellow butterflies flitting past and carried on the breeze, the laughing call of the green woodpecker.

The grazing management of the chalk grassland is integral to the farming system and enables over one hundred plant species to flourish, free from encroaching scrub. Chris has entered the grassland into a government "agri-environment" scheme agreement under which he receives $90 per hectare (at a conversion rate of $1.50 to 1 pound). The payment is in return for his continued grazing management without the use of fertilizers, herbicides, and pesticides. The level of payment is calculated on the basis of the "income foregone" by farming with these restrictions.

Applesham Farm is within an agri-environment scheme called the South Downs Environmentally Sensitive Area (ESA), which covers 69,045 hectares of the counties of East Sussex, West Sussex, and Hampshire. The South Downs was one of the first ESAs to be designated under the Agriculture Act of 1986, which gave the Minister for Agriculture power to designate land and make payments to farmers in areas of national environmental significance. The Act was the democratic response to the widespread desire to protect nature on farmland. ESA payments cover landscape and historic features as well as nature; for example, in the South Downs ESA, money is given for renovation of traditional farm buildings and for maintaining permanent grassland to protect archeological remains from the plow. Finance is also available for habitat creation. Chris has entered 35 hectares of steeper, less-fertile land into an option for conversion of arable to permanent grassland managed in a low-intensity way at a payment rate of $360 per hectare. During our visit, we found that plants such as bird's foot trefoil and quaking grass, which are characteristic of chalk grassland, were already showing up in the reverted grassland. These discoveries were very encouraging, though we do not expect a rapid return to a rich flora. Research suggests that at least a century may be needed for a grassland on former arable land to develop a composition that is similar to ancient grassland (Gibson and Brown 1991).

There are now forty-three ESAs in the United Kingdom of England, Scotland, Wales, and Northern Ireland, and they cover 2.5 million hectares in total, which is about 15 percent of the farmed land. There are twenty-two ESAs covering 0.94 million hectares in England, including such diverse landscapes as the Somerset Levels and Moors, the Cotswold Hills, the Pennine Dales, and the Upper Thames Tributaries. To provide some idea of what these areas mean on the scale of the United States, the United Kingdom (commonly referred to as Britain) is similar in size to Illinois and Indiana combined, while the area of England is close to that of Alabama. Uptake of the schemes by farmers has been good. For example in the South Downs ESA, 5,566 hectares has been entered into the chalk grassland option and 5,275 hectares into the reversion of arable to grassland option (Ministry of Agriculture Fisheries and Food unpublished data, March 2000). To date, 530,000 hectares have been entered into ESA agreements in England.

The ESAs are not the only government agri-environment schemes available to farmers. In 1991, the Countryside Stewardship Scheme was launched in England. The aims and options are similar to those of ESAs but agreements are not limited to particular geographical areas of the country but are made over a series of habitats or landscape types, such as chalk and limestone landscapes and old meadows and pastures, wherever they occur in England. Similar schemes operate in the rest of the United Kingdom. By the year 2000, agreements in England covered 209,609 hectares. Government expenditure on both of these agri-environment schemes totaled $108 million in England in 1999/2000 (Ministry of Agriculture Fisheries and Food 2000a).

Agri-environment schemes have worked because of the financial incentives. As Chris emphasized to us, farmers could not afford to undertake these conservation measures without the agri-environment payments. Entry into schemes is entirely voluntary on the farmer's part. There is no long-term commitment beyond the first ten-year agreement, but so far little land has been taken out of schemes. Changes in land ownership have triggered withdrawal. In one case in 1997 on the South Downs, the new owner plowed land previously under agreement, which meant that $375,000 had been spent to no avail.

Based on the results of monitoring the effects of the schemes, the management regimes specified by the various options are being refined to increase their effectiveness, especially for individual species. For instance, the rare marsh fritillary butterfly is suited by light grazing that produces a grassland around 8 to 20 centimeters in height.

As well as the introduction of voluntary agri-environment schemes, the regulatory framework has been strengthened for the best wildlife sites. These are designated Sites of Special Scientific Interest (SSSI) under the Wildlife and Countryside Act of 1981. The chalk grassland on Chanctonbury Hill is now within an SSSI, which means that farmers must consult the statutory nature conservation agency, English Nature, if they want to agriculturally improve the land. English Nature can also make payments under management agreements to enable low-intensity methods of farming to be followed and thereby protect the wild plants and animals found on these sites. There are over four thousand SSSIs in England covering the full range of habitats from coast to mountaintop and including important geological features. Within the SSSI series there are 304 chalk grassland SSSIs totaling 23,600 hectares (English Nature, unpublished data, 1996).

Although agri-environment schemes and the SSSI series have been very valuable in maintaining nature on farmland, there is worrying evidence of continuing declines in wildlife across the countryside in general. The national declines in farmland birds have attracted most attention. Volunteer birders have been systematically censusing farmland birds in sample plots in Britain since the early 1960s. They have found that between 1970 and 1990 there have been marked declines in the populations and ranges of many farmland birds (Fuller et al. 1995). Seven species were estimated to have undergone population decreases of at least 50 percent, including formerly common and widespread species such as the corn bunting (down 76 percent), grey partridge (down 73 percent) and skylark (down 54 percent).

Farmland birds have maintained good populations on Chris Passmore's land. For instance, in the spring of 2000, a volunteer birder recorded twenty-seven singing male skylarks in an area of about 1 square kilometer. This number of birds represents twenty-seven occupied territories and is six times the average density of four and one-half per square kilometer in the general countryside (Browne et al. 2000). In fact, Applesham Farm has been at the center of research into the factors controlling farmland bird numbers. The grey partridge and corn bunting have been studied in detail on Chris's farm and neighboring farms by researchers from the Game Conservancy Trust and Sussex University. We saw a clue to the puzzle on our visit to the farm while we sat looking across the coomb at the pattern of fields laid out before us. At first glance the landscape looks much the same as the surrounding downs. However, Chris pointed out the much greater variety of color tints on his land. Instead of the uniform pale chalky brown of the plowed-land characteristic of surrounding farmland, his fields were a patchwork of golden wheat and barley stubble, light, bright green of undersown grass and clover, now revealed by the harvest of the accompanying grain crop and the paler green of older sown leys, all bound together by lines of bleached, seeding grasses marking the fence lines and farm lanes. Chris explained that the pattern is there because Applesham is a mixed farm, unlike the continuous cropping enterprises of surrounding farms. He has found that the rotation of grass and grain crops and the rearing of livestock makes sound commercial sense on his farm and at the same time benefits wildlife in several ways.

The grass and clover ley breaks arable pest and disease cycles while the arable period destroys intestinal parasites of livestock so worming treatments are unnecessary. The nitrogen-fixing action of the clover in the undersown ley means that no inorganic nitrogen is required for good grass growth, and it reduces the amount of inorganic nitrogen that needs to be applied to the arable crops. Research has shown that the rotation also enables invertebrates that are important food for farmland birds to complete their life cycle. Sawfly caterpillars are a particularly favored food for grey partridge, corn bunting, and skylark. Sawfly eggs are laid in quantity in spring-planted barley and undersown fields. Caterpillars that escape being eaten overwinter as pupae in the soil and then emerge in April and May. They survive better where no plowing occurs during the winter period, as is the case for undersown fields (Potts 1997, Barker et al. 1999).

The stubbles left after harvest at Applesham Farm are not plowed immediately because they are grazed by sheep and cattle, which spend the winter outside. Farmland birds feed on the remnant grain and on weed seeds in the stubble. The hay and straw spread across the fields as supplementary livestock feed in winter also provide feeding stations for seed-eating corn bunting, linnets, and yellowhammers.

The tall, uncropped grassland along fence lines and farm lanes provides reservoirs of invertebrates and are popularly known as beetle banks. The value of beetle banks for pest control has been well researched (Potts 1997). Chris has not had to spray his grain crops during the summer to control aphids for thirty years because the natural invertebrate predators harbored by the fence lines, undersown leys, and chalk grassland keep aphid numbers below the economic threshold for spraying. The rough grassland at Applesham Farm also provides habitat for small mammals that are preyed upon by the tawny and barn owls that nest in Chris's farm buildings.

The Future of Farming and Nature

The importance of mixed farming and uncropped land around fields for maintaining nature on farms is beginning to be understood and acted upon by government. The South Downs ESA now includes options for undersown arable crops, winter stubbles, and uncropped field margins. Similar options for arable land have been included in a pilot Arable Stewardship Scheme in East Anglia and the West Midlands and are likely to be available more widely in future. Overall, agri-environment spending in England is planned to increase to $261 million a year by 2006 (Ministry of Agriculture Fisheries and Food 2000a) to the benefit of the whole range of farmland habitats and species.

The profile of farmland wildlife has been raised by the publication of the *U.K. Biodiversity Action Plan* as part of the U.K. government's response to the Biodiversity Convention signed in Rio de Janeiro in 1992 (United Kingdom Government 1994). Individual habitat and species action plans have since been produced that set tough conservation targets, including, for example, stopping further loss of unimproved grasslands (U.K. Biodiversity Group 1998). Another example is the grey partridge action plan, which has targets to halt population declines by 2005 and ensure the population is above 150,000 pairs by 2010 (U.K. Steering Group 1995). Action to create habitats to expand existing fragments and join isolated sites is also included in the plans, such as the creation of 6,000 hectares of heathland by 2005 (U.K. Steering Group 1995). In a wider policy context, the U.K. government has adopted an index of abundance of farmland birds as a biodiversity indicator for both sustainable agriculture (Ministry of Agriculture Fisheries and Food 2000b) and the general quality of life in the United Kingdom (Department of the Environment Transport and the Regions 1999).

Consumer pressure is an increasingly important factor in influencing farming practice. The demand for organically grown food has risen sharply in recent years and the amount of land managed organically or being converted to organic management has expanded as a result. The area doubled between 1998 and 1999 to 240,000 hectares in the United Kingdom. As yet organic farming forms only a small proportion of the agricultural industry and 70 percent of organic produce has to be imported from abroad to satisfy demand (Stopes et al. 1999). Consumer interest in organic food probably relates more to concerns about the safety of food after a number of high-profile health scares such as the Bovine Spongiform Encephalopathy cattle disease crisis, which hit the livestock industry in 1990s, rather than perceived environmental benefits. However, research suggests that wildlife does benefit from organic farming compared to conventional farming. For example, skylark densities have been found to be higher on organically cropped fields than conventionally managed fields (Wilson et al. 1997).

All of these developments are hopeful signs that the place of nature on farms in Britain will be maintained and restored in the twenty-first century. However, the fickleness of farming and wildlife fortunes is well illustrated by the recent outbreak of Hoof and Mouth Disease in Britain. The disease struck in February 2001 and by June 2001 over 4 million animals had been slaughtered in the attempt to control the spread of the disease. This represents 7% of the national herd. It is too soon to say what the long-term effects on farmland nature will be. A pessimistic view is that livestock farming in the lowlands will be further reduced, resulting in less mixed farming and greater difficulties in obtaining grazing animals for grasslands of high nature conservation value. An optimistic view is that the fundamental review of agricultural policy promised by the government in the wake of the disease will result in strong growth of environmentally sustainable farming.

There are some large obstacles ahead in terms of reforming the Common Agricultural Policy away from production subsidies toward environmental rewards for farmers and in gaining recognition in the world trade arena for the multifunctional character of farmland. Just as in previous times when the land

yielded products for a multiplicity of uses so farmland now can not only produce food in a sustainable way but also be a place for nature and the enjoyment of nature by people.

The biodiversity of farmland would have a more secure future if its value could be separated from income foregone and production-related calculations. Skylarks are not traded on the world commodity markets and have nonnegotiable habitat requirements. The case for environmental rewards for farmers needs to be made at the international level, although the development of a scale of values for biodiversity that is independent of the economics of commodity production is some way off. If these fundamental policy reforms go ahead, they hold the promise that farmland in Britain and elsewhere can produce high-quality food and a rich and satisfying landscape far into the future.

Acknowledgements

The authors thank Chris Passmore for a wonderful and informative day at Applesham Farm, Duncan Coe of Wiltshire County Council for helpful information on Barbury Hill Fort and the prehistory of Wiltshire, and our colleagues Phil Grice and Karen Mitchell for useful discussion and information on farmland birds and agricultural policy.

References

Allen, D. E. 1976. *The Naturalist in Britain: A Social History*. Allen Lane, London.

Barker, A. M., N. J. Brown, and C. J. M. Reynolds. 1999. "Do Host-Plant Requirements and Mortality from Soil Cultivation Determine the Distribution of Graminivorous Sawflies on Farmland?" *Journal of Applied Ecology* 36:271–282.

Browne, S., J. Vickery, and D. Chamberlain. 2000. "Densities and Population Estimates of Breeding Skylarks *Alauda arvensis* in Britain in 1997." *Bird Study* 47:52–65.

Department of the Environment, Transport, and the Regions. 1999. *Quality of Life Counts: Indicators for a Strategy for Sustainable Development for the United Kingdom: A Baseline Assessment*. Her Majesty's Stationary Office, London.

Duffey, E., M. G. Morris, J. Sheail, L. K. Ward, D. A. Wells, and T. C. E. Wells. 1974. *Grassland Ecology and Wildlife Management*. Chapman and Hall, London.

Fuller, R. J., R. D. Gregory, D. W. Gibbons, J. H. Marchant, J. D. Wilson, S. R. Baillie, and N. Carter. 1995. "Population Declines and Range Contractions among Lowland Farmland Birds in Britain." *Conservation Biology* 9:1425–1441.

Fuller, R. M. 1987. "The Changing Extent and Conservation Interest of Lowland Grasslands in England and Wales: A Review of Grassland Surveys 1930–1984." *Biological Conservation* 40:281–300.

Gibson, C. W. D., and V. K. Brown. 1991. "The Nature and Rate of Development of Calcareous Grasslands in Southern England." *Biological Conservation* 58:297–316.

Hoskins, W. G. 1955. *The Making of the English Landscape*. Hodder & Stoughton, London.

Lambrick, G., and M. Robinson. 1988. "The Development of Floodplain Grassland in the Upper Thames Valley." Pp. 55–75 in *Archeology and the Flora of the British*

Isles, edited by M. Jones. Botanical Society of the British Isles Conference Report No 19. University of Oxford Committee for Archaeology, Oxford.
Mabey, R. 1996. *Flora Britannica.* Sinclair-Stevenson, London.
Ministry of Agriculture, Fisheries, and Food. 2000a. *England Rural Development Plan 2000–2006.* Ministry of Agriculture, Fisheries, and Food, London.
——. 2000b. *Towards Sustainable Agriculture: A Pilot Set of Indicators.* Ministry of Agriculture, Fisheries, and Food, London.
Nature Conservancy Council. 1984. *Nature Conservation in Great Britain.* Nature Conservancy Council, Peterborough, U.K.
O'Connor, R. J., and M. Shrubb. 1986. *Farming and Birds.* Cambridge University Press, Cambridge.
Payne, K. R. 1998. *Management Plan for North Meadow National Nature Reserve.* Nature Conservancy Council for England (English Nature), Devizes, U.K.
Perring, F. H., and S. M. Walters. 1962. *Atlas of the British Flora.* Thomas Nelson & Sons, London.
Potts, D. 1997. "Cereal Farming, Pesticides, and Grey Partridges." Pp. 150–177 in *Farming and Birds in Europe: The Common Agricultural Policy and Its Implications for Bird Conservation*, edited by D. J. Pain and M. W. Pienkowski. Academic Press, London.
Rackham, O. 1986. *The History of the Countryside.* J. M. Dent & Sons, London.
Shoard, M. 1980. *The Theft of the Countryside.* Temple Smith, London.
Stopes, C., M. Redman, and D. Harrison. 1999. *The Organic Farming Environment.* The Soil Association, Bristol, U.K.
U.K. Biodiversity Group. 1998. *Tranche 2 Action Plans.* Vol. 2, *Terrestrial and Freshwater Species and Habitats.* English Nature/U.K. Biodiversity Group, Peterborough, U.K.
U.K. Steering Group. 1995. *Biodiversity: The U.K. Steering Group Report.* Her Majesty's Stationary Office, London.
United Kingdom Government. 1994. *Biodiversity: The United Kingdom Action Plan.* Her Majesty's Stationary Office, London.
Wigginton, M. J., ed. 1999.*Vascular Plants.* 3rd ed. Vol.1 of *British Red Data Books.* Joint Nature Conservation Committee, Peterborough, U.K.
Wilson, J. D., J. Evans, S. J. Browne, and J. R. King. 1997. "Territory Distribution and Breeding Success of Skylarks (*Alauda arvensis*) on Organic and Intensive Farmland in Southern England." *Journal of Applied Ecology* 34:1462–1478.
Woodward, M. 1994. *Gerard's Herbal: The History of Plants.* Senate, London.

Chapter 10

Restoring Prairie Processes to Farmlands

Laura L. Jackson

If conservation biologists could have anything they wanted, what would agriculture look like? Recognizing that land use is shaped by policy, values, culture, economics, and physical infrastructure, in addition to biological parameters such as climate and soils, it nevertheless is the responsibility of conservation biologists to think about what strategies would be most likely to preserve biological diversity (and people's enjoyment of it) while still supplying us with our needs. There is a good deal of territory in between "wilderness" and "agricultural wasteland." We will not know just how much yield or how much biological diversity need be compromised until we dream up these strategies, try them out, fine tune them, and incorporate them into our culture and economy.

The book lays out a vision of restoring biological diversity and ecosystem services to areas that have been damaged by intensive industrial agriculture. In this chapter, I will describe some of the ecosystem processes and structures that were altered, first by the great conversion of prairie to agriculture in the upper Midwest and then by the conversion of traditional rotation-based cropping systems to corn-soybean monocultures. The essential differences between these agroecosystems—how and why they have changed, and the consequences of those changes—can shed light on how we might proceed, practically, to recognize and restore important ecosystem functions within working agricultural landscapes.

Land Use Change and Ecosystem Processes in the Upper Midwest

When I first moved to the Upper Midwest, I wanted to learn more about the details of the great plowdown that had converted the prairies to the farms I saw around me. According to most firsthand witnesses and prairie ecologists, the prairie had virtually vanished by 1900 (MacBride 1895, Quick 1925, Smith 1992). However, students of natural history provided some evidence to the contrary. *Pastures and Meadows of Iowa*, published in 1901 (Pammel et al. 1901), described a wealth of native grasses and wildflowers in Iowa's pastures and hay meadows. Bohumil Shimek (1917) documented rapid reinvasion of abandoned agricultural fields, disturbed road cuts, and railroad rights-of-way by native prairie plants. Seed sources were "amply sufficient" in his view to reseed all suitable areas—meaning there were reservoirs of native plants everywhere. Today, disturbed sites are instead immediately recolonized by weedy, exotic species. In search of large prairie tracts in the 1940s, Ada Hayden (1947) found the countryside chock full of prairie along roadsides and railroad rights-of-way, in pioneer cemeteries, and numerous other nooks and crannies of the landscape. Finally, many of the dried specimens of prairie plants at the University of Northern Iowa herbarium (where plants collected in the field are dried flat, glued to large sheets of paper, and stored under cool, dry conditions) had been collected *on farms* as late as the 1970s.

What was the nature of Iowa's agriculture before 1970, that it could harbor such wild diversity? I turned to state agricultural statistics published starting in the 1860s. Between 1860 and 1900, total land in farms increased from 2.7 million acres to 34.6 million acres, mainly at the expense of prairies. To my surprise, corn acreage did not increase steadily during that period but wobbled back and forth between 8.4 million and 11.7 million acres, and soybeans did not occupy any major acreage until after World War II. Significant areas of prairie were cut for hay well into the 1960s. "Tame hay" was also a major part of the landscape, so important it was divided into several categories, including varying proportions of timothy, bluegrass, and clover. A diversity of small grains, including spring and winter wheat, rye, barley, flax, buckwheat, and oats made up 20 percent of the land in farms in the late 1930s. The statisticians also reported production figures for orchards, vineyards, vegetables, hops, honey, and potatoes (Iowa Executive Council 1854–1925, Iowa State Agricultural Society 1865–1899, Iowa Department of Agriculture 1900–1985, Iowa Agricultural Statistics 1986–1990, NASS 2000).

The statistics did not come into focus for me until I visited a small organic dairy farm in Black Hawk County, Iowa, run by brothers Richard and Albert Steffen. The Steffens stopped using inorganic fertilizers and pesticides in 1965 and reinstituted a five-year crop rotation similar to what they had grown up with in the 1930s. Their crop rotation consisted of oats undersown with red clover the first year, a hay crop of clover and timothy grass the second year, corn the third year, soybeans the fourth year, and a second crop of corn the fifth year. Dairy and beef cattle ate most of the crops they produced and provided the farm income. On the surface, this didn't seem so different from a "normal" Iowa farm.

Patiently, the Steffens explained how they managed without fertilizers and herbicides. Two years of sod crops (oats, hay) helped to control the weeds that flourished in the annual row crops (corn, beans). Clean cultivation of the row crops (corn, beans), seeded in late spring when the soil warmed, controlled the cool-season weeds that thrived under oat cultivation as well as the perennial weeds that appeared during the sod crop years. Soil organic matter, crumb structure, and tilth waxed under the sod and waned under the cultivated crops. Clover hay and cattle manure provided nitrogen for corn production. They purchased a small amount of pelletized chicken manure to fertilize the second crop of corn. This system had kept the Steffens in business for thirty years without benefit of a price premium for organic milk and meat.

After learning this, I went back to the data and aggregated individual crop data into row crops versus sod crops (Figure 10-1). Now the essential pattern stood out: sod and row crops had been in a fifty-fifty balance statewide from the 1860s to the 1950s. This seemed remarkable, given the changes in economics, technology, and farm policy over this period and the wide variety of crops farmers could grow in these good prairie soils. It suggested to me that there were more fundamental, ecological forces keeping the sod crop–row crop system in place. Beginning fitfully in 1940 and steadily after 1957—after tractors had replaced horses—soybeans began to take the place of the sod crops and pastures, transforming the Iowa landscape (Figures 10-1 and 10-2).

By the 1970s, the transition was in full swing. Agricultural statisticians stopped reporting "minor crops" altogether and began rounding their numbers to the nearest 10,000 acres. As crops became less diverse, livestock followed suit. Family-sized poultry operations began to disappear as the processors moved to the southeastern United States. At roughly the same time, huge beef feedlots cropped up on the High Plains, where cheap underground water resources fed irrigated corn. Beef processors soon followed. Facing lower prices and reduced options for marketing their cattle, many Iowa farmers got out of the beef cattle business, with dairy following soon after. Hogs, long the "mortgage burner" for the family farm, remained firmly in the hands of independent producers until the late 1980s. Unlike poultry, beef, and dairy production, hogs remained an important part of the agroecosystem—they just moved to corporate controlled factory-scale operations. In 1987, Iowa had 36,670 hog producers. In 1999, there were 14,500 hog producers, raising roughly the same number of hogs.

The original plowdown from the 1850s to the 1890s eradicated the prairie plants and animals with large area requirements (Smith 1992). However, traditional rotation-based cropping systems maintained important elements of the prairie ecosystem: a grassy landscape of diverse, perennial hayfields and

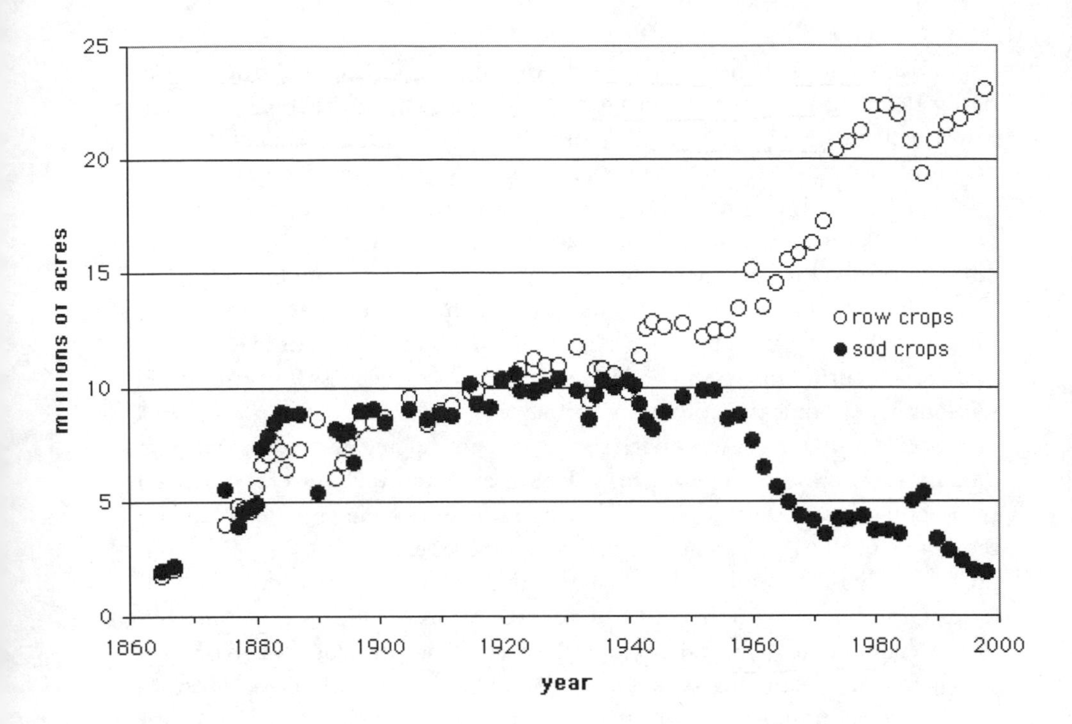

Fig. 10-1. (Above) Change in the relative amounts of row crops (corn and soybeans) and sod crops (crops planted close together, such as small grains, and hay) from 1865 to 1999, based on Iowa agricultural statistics (see text for references). Adapted from Jackson (1998).

Fig. 10-2. (Right) Simulated map of prairie hay, tame hay, pasture, and other agricultural grasses in Bremer County, Iowa, illustrating changes in the landscape over time. Each square represents a field. Positions of individual fields were randomly assigned. The maps are created from the following data, based on county agricultural statistics (see text for references). In 1910, prairie hay made up 8.8 percent of land in farms, tame hay 8.5 percent, and pasture 29 percent. In 1940, prairie hay made up 3.1 percent, tame hay 8.4 percent, and pasture 30.8 percent. In 1970, prairie hay made up an estimated 0.1 percent, tame hay 8.3 percent, and pasture 17.9 percent. In 1997 (the last date that pasture data were available), prairie hay made up 0 percent, tame hay 4.3 percent, pasture 5.3 percent, and conservation reserve and cover crops made up 3.5 percent of the total land in farms. The real spatial distribution of land uses would probably have been more clustered with prairie hay and pasture mainly in poorly drained or rocky soils.

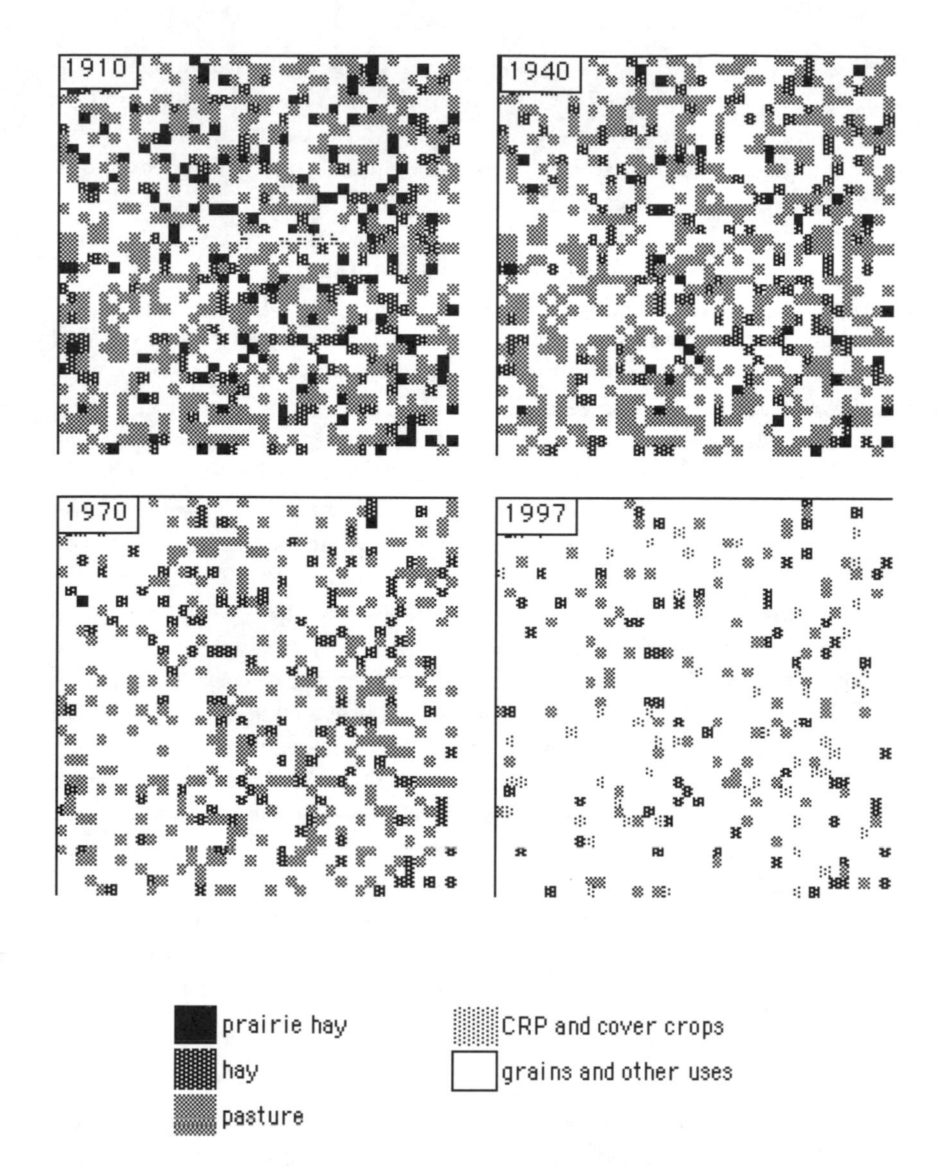
1910
1940
1970
1997
prairie hay
hay
pasture
CRP and cover crops
grains and other uses

pastures intermixed with fields of annual grain crops; biological nitrogen fixation by legumes; well-timed delivery of nitrogen to plants via soil organic matter mineralization; and an intact food chain, including ruminant herbivores. How did the change to row crop monocultures and confined livestock affect these ecosystem processes?

A Monoculture of Annual Plants

Prairies are covered with a diverse mixture of perennial plant species. Perennial means that the plants come up from roots year after year instead of from seed, and as a consequence the physical structure of the vegetation—dry plant litter, crowns, and roots—protects the soil from erosion. Because of the diversity of species growing and maturing at different times, prairies harvest solar energy over the entire growing season, from early May until well after frost in October. Under diverse crop rotations about half of the landscape was covered by perennial plants (hay) or closely spaced, sod-forming annuals (oats or other small grains). These protected the soil and slowed the movement of water (Figure 10-2). The diversity of crops allowed them to harvest solar energy over the entire growing season.

After the second plowdown, the agroecosystem involved just two species of annual plants, plus a few weeds. Corn, a tall, heat-loving grass, does share the same method of photosynthesis as the tall prairie grasses, and soybeans do fix some nitrogen, but the similarities with prairie end there. Because these crops are planted in late spring and are slow to cover the ground, satellite images of vegetation "greenness" provided by the National Agricultural Statistics Service color the Upper Midwest in yellows and tans as late as mid-June. Likewise, this region begins to turn yellowish-brown by mid-September, when there are still plenty of warm days and soil moisture. The corn-bean agroecosystem forfeits about two months worth of photosynthesis (turning sunshine into food—the ultimate ecosystem service). The soils under corn and soybeans are virtually bare for seven months out of the year.

Prior to the conversion of prairie to crop rotations, rainfall and snowmelt flowed slowly through the sponge of leaves and roots. Upland prairie streams flowed long after the rains had come and gone, as the soil slowly released water that had not been taken up by the vegetation. After agricultural development, raindrops encountered plowed fields, hayfields, and closely cropped pastures. Runoff rates were faster and upland streams changed forever. Herbert Quick (1925) notes that

> All those beautiful brooks . . . were like delicate flowers, too tender for the touch of humanity. In those old days the water of the rains

> flowed freely down the slopes and into the sloughs to seep gradually into the brooks. Then came the human flood and turned those slopes into plowed farms. . . . [T]he waters of the brooks gradually dried up save in wet weather. . . . (74–75)

The great increase in row crops after 1957, the increased use of subsurface drainage, and the general reduction in soil organic matter over time has exacerbated this problem. As flood peaks increased, we responded by channelizing and straightening streams. In 1975, it was estimated that as much as 3,000 miles of Iowa's streams and small rivers had been eliminated by straightening (Bulkley 1975).

Beyond Fragmentation: The Homogenized Landscape

Some explorers found the prairie landscape flat, monotonous, even "barren." The grasses waved for miles in the distance, broken only by a few willows along the streams and the occasional stand of burr oaks. In contrast, the era of diverse crop rotations created a more diverse patchwork of agricultural grasslands, bare ground, grain crops, prairie hay, and wooded homesteads. Not only did grassland nesting species still breed in the hay meadows and pastures of these early farms, but birds of forests could inhabit the farmsteads while birds of open ground, such as the killdeer, could use the farmed ground (Best et al. 1995). We had lost the *large* areas of wilderness in the Upper Midwest (MacBride 1895) but long crop rotations were still able to support many smaller prairie and woodland species.

The second plowdown homogenized this landscape, turning already fragmented remnants into increasingly smaller and more isolated pieces. Soybeans ate up the patchwork landscape. With less crop diversity, every field needed the same field operations (plow, disk, plant, cultivate, fertilize, spray, harvest) at precisely the same time, and speed became critical. Fields got bigger in lockstep with machinery. Fencerows and small patches of rocky, wet, or otherwise marginal ground took too much time to go around. One farmer told me about paying for ninety dump-truck loads of fill dirt for a sinkhole (a depression created by the collapse of a bridge of limestone). It was expensive, he said, but worth it. Such "improvements" still go on today, seemingly without regard to whether the crops planted there could ever pay for themselves. One dry spring a few years ago I watched a farmer bulldoze the willows along a sizeable backwater channel of the Cedar River that cut his cornfield in half. During years of normal rainfall the channel floods, erodes more soil, and his corn crop fails.

Bison and elk had been the principal herbivores on the tallgrass prairie, at least at the time Europeans arrived. Cattle took their place on the diversified farm. Without cattle, there would be no reason for perennial grasses on the farm. (The ruminant gut changes "worthless" cellulose into cash.) Although cattle eliminated the grazing-intolerant native species, their presence kept the landscape perennial (see Figure 10-2) and protected much marginal land from the plow. Rotting barns once used for storing hay, and drooping barbed wire fences tell not simply of rural depopulation, but also of lost habitat, and, as I will discuss below, nutrient cycling.

So the diversity of wildlife habitats on farms ratchets downward: first the large mammals and birds, and then those with smaller and smaller area requirements (Dinsmore 1994). According to eighty years worth of data from the National Breeding Bird Survey (Jackson et al. 1996), grassland nesting birds have been in decline since the 1960s. (Smaller mammals, plus amphibians, reptiles, and invertebrates may be just as sensitive to these changes in landscape pattern, but, unfortunately, there are no comparable long-term data sets to track them.) It seems absurd, but now even those animals with the most modest area requirements—plants and the insects that pollinate them—are running out of viable places to live, and—just as important—avenues of movement. All organisms need to be able to move and disperse and to find new territories and new mates so their offspring can adapt to new conditions (especially with global warming coming on). The homogenization of the landscape threatens their ability to move. For instance, plants move via their pollen and seeds. The federally threatened eastern prairie fringed orchid is pollinated by sphinx moths, which in some cases would have to cross many miles of insecticide-treated fields to cross-pollinate these orchids.

Nutrient Dynamics and Soil Health

On the prairie, wild legumes provided most of the plant-available nitrogen *via* biological nitrogen fixation (the plant converts solar energy to food for *Rhizobia* bacteria living in its roots, and the bacteria convert atmospheric nitrogen into a form available to plants). In the era of diverse crop rotations, legumes were still the main source of nitrogen. The switch to corn-beans agriculture had far-reaching implications for fertilizer use, because soybeans do not provide nitrogen above the amount they consume. Recently, researchers have estimated that humans have doubled the rate of nitrogen input into the terrestrial nitrogen cycle, and most of this has happened since 1975 due to industrially produced nitrogen fertilizer (Vitousek et al. 1997).

On the prairie and the diversified farm, plants acquired their nutrients primarily from decaying organic matter (plant litter and animal manures). Nutrients in organic matter became available gradually, as the soil warmed up, and in rough parallel to the rise in plant nutrient demand. This minimized nitrogen losses, because the nutrients were removed by plants just as they were made available. In the corn-bean system, plants acquire their nutrients mainly from free nitrates in the soil. The most commonly used nitrogen fertilizer (anhydrous ammonia) becomes the highly mobile nitrate as soon as soil temperatures rise above 50 degrees Fahrenheit. The biological processes that linked nutrient availability and crop uptake are no longer operating, and therefore industrially fixed nitrogen is vulnerable to leaching long before crop plants can consume it. According to some accounts, only about half of the nitrogen fertilizer applied in the Upper Midwest is taken up by crops.

Adding livestock and grain crops to the prairie certainly changed nutrient dynamics, but there remained an intact food chain on practically every farm, with herbaceous perennials at the base, large herbivores in the middle, predators (humans) at the top, and animal waste everywhere. Conversion to chemically fertilized row crops has increased nitrate pollution of surface waters. Randall (Randall et al. 1997, Randall 2000) has found that nitrate-nitrogen losses (nitrogen is converted to nitrate by soil bacteria before it enters waterways) are thirty to fifty times higher than from land planted to perennial hay crops or in perennial grass systems.

Now many farms raise grains only, and livestock (especially hogs, poultry, and dairy cattle) are raised in a few very large facilities called *confined animal feeding operations* (CAFOs). Nutrient dynamics, as a consequence, have become even more distorted. Some cash grain farms haven't had animal manure applied to their fields in twenty years; meanwhile, others are flooded with it. In a study of hyperconcentrated swine in north-central Iowa (60,000 finishing hogs housed within a 2-mile radius) we found that the manure was spread on only 2,450 crop acres. Some soybean fields received 4,000 to 6,000 gallons of manure per acre even though soybeans do not need extra nitrogen. We found that the CAFOs would need three times as much land to apply manure efficiently for its nitrogen content, and ten times as much land for the phosphorus content (Jackson et al. 2000).

Perennializing the Agricultural Landscape

Many authors have proposed that we consciously mimic aspects of nature to develop new crops, crop assemblages, or farming systems to take advantage of intrinsic properties in natural systems (Lefroy et al. 1999). Jackson (1980, 1987) and Soule and Piper (1992) reasoned that if fields of annual grain crops such as corn and wheat could be replaced by mixtures of perennial plants that did not have to be replanted every year, many of the protective aspects of the prairie could be revived. Herbaceous perennial plants, grown in multispecies mixtures as they do in the prairie, would sponsor their own nitrogen fertility with nitrogen-fixing legumes, would hold the soil year-round, reduce insect and weed pressure compared to monocultures, and reduce dependence on fuel-intensive tillage. This idea, called *natural systems agriculture*, strikes at the heart of many agricultural problems and addresses their underlying causes.

New perennial crops will require decades of plant breeding and agronomic research. Domestication of new crops from wild species is quite rare; all of our grains and legumes were domesticated several thousand years ago under conditions that are still not well understood. We have little experience selecting for high yield in herbaceous perennials, although the yields of various domesticated woody perennials are encouraging. There are a number of agronomic and cultural problems to work out—how plant species composition would change over several years; how weeds and herbivorous insects would be managed; how nutrient exported at harvest could be replaced; and (not least of the concerns) how these new species would be incorporated into human and livestock diets. Recent plant breeding efforts at the University of Washington and at The Land Institute have focused on perennializing cereal grains. This approach may prove faster than domesticating wild species but will still need legumes and a variety of other species to more fully mimic a prairie plant community.

Even if perennial grain crops become available, it is a good bet that we will still be dealing with the problems of annual crops like corn and soybeans long in to the future. Conservation biologists need to ask, are there any *other* ways to restore the agricultural landscape with the crop species we have now? Perennializing grain crops may be a long-term goal, but we could begin now to "perennialize" the landscape.

Although it doesn't sound radical enough to solve all agricultural problems, a return to diverse crop rotations with sod-forming crops in two or three of every five years would be a giant step in the right direction. If farm policy favored such a change, we would have the crop plants, livestock, equipment, buildings, and other infrastructure, and most importantly the cultural knowledge to implement this within a decade. Crop rotations involving small grains and legume-grass hay mixtures have a number of proven benefits to soil health (Logsdon et al. 1993) and their use could eliminate or drastically reduce the need for the 45 million pounds of herbicides and insecticides, and the 1.5 billion pounds of nitrogen fertilizer applied annually to Iowa fields alone (NASS 2000).

Innovations within Old Patterns

Although I am suggesting we "go back" in a sense to cropping systems in use prior to the 1950s, that does not necessarily mean that we will "go back" to backbreaking labor, horses, or straw hats. Both older and recent innovations will prove crucial. For instance, farming on the contour and strip cropping will still be necessary in even moderately hilly terrain for preventing soil erosion during the row crop part of the rotation. Innovative tillage systems and implements such as ridge tillage, designed to minimize soil disturbance while maximizing weed control, offer farmers better options than they had in the past. And thanks to new kinds of machinery, hay production and storage no longer require as much labor as they once did.

Many innovations are possible within the basic small-grains–hay–row-crop rotation. For instance, in the small-grains–hay year of his rotation, when many farmers would plant only oats and clover, Mike Natvig plants ten species: oats, wheat, barley, and field peas, three kinds of clover, alfalfa, timothy, and orchardgrass. The field peas climb up the stalks of the small grains and all are harvested together using a combine (the resulting high-protein "succotash" is fed to pigs). As the grains are harvested, the six-species hay mixture is already sprouting. Natvig believes that the species diversity helps captures a larger proportion of the year's solar energy as different plants grow and mature at different times, and that it ensures a better crop over varying field conditions.

Cover crops and green manures such as hairy vetch, planted before or after a cash crop, help to restore the diversity of plant forms and growth patterns characteristic of prairies, in order to fix nitrogen, protect the soil, and suppress weeds. There are numerous opportunities to adapt old cover crops and to develop new ones. Another area in which more research is needed is that of soil fertility on farms using complex crop rotations. With funding from the Leopold Center for Sustainable Agriculture, the Michael Fields Agricultural Institute in Wisconsin is currently refining a soil carbon model to do this. Farmers will be able to estimate how much nitrogen will be available to the next year's crop from decomposition of residue (roots, stems) of previous years' crops. This should reduce the amount of excess nitrogen applied as fertilizer.

Rotational grazing is another innovation not practiced by farmers in the 1950s. Improvements in portable electric fencing and watering devices have made it possible to move cattle frequently and thus increase pasture productivity and profits. This may not look like ecological restoration, but in chapter 6, Brian DeVore describes how rotational grazing has been used to heal stream banks ruined by decades of continuous overgrazing. It's also restoration from the point of view of perennial cover and bird species. In 1994, Matt and Diana Stewart took 200 acres out of corn and soybeans, and planted pasture grasses and clovers to feed their dairy herd. This rivals the size of eastern Iowa's largest tallgrass prairie, Hayden Prairie, which is only 240 acres. Hurley O'Hara (1999) compared bird incidence and abundance in the Stewart pasture and found six of the eight grassland nesting species found at Hayden

Prairie. While these pastures may not look like prairie restoration to a botanist, they apparently do to a bobolink.

Restoration of Nutrient Dynamics and a Dispersed Food Chain

Return to crop rotations integrated with livestock would help to restore the synchrony between nitrogen availability and nitrogen uptake found in prairie soils and would again distribute animal waste more equitably on the landscape. If we persist in growing high-nitrogen-demanding grains such as corn, the safest way to do so is by accumulating nitrogen in the form of soil organic matter (from manure and grass-legume hay) and making it available through decomposition throughout the growing season. Corn breeders may have to select for varieties that work better with this source of fertility.

Why would it be preferable to have livestock spread sparsely across the landscape rather than concentrated in a few central locations as they are now? Proponents of the concentrated livestock industry, for instance, assert that there is no intrinsic advantage to decentralizing operations and that a few large operations are easier to regulate than thousands of small ones. To answer this question, Terrance Loecke and I set out to quantify the nutrient efficiency of eight small, diversified crop-livestock operations and compare them to factory-style swine operations. A high net nutrient import (imports in feed minus exports in meat) means that a farm is using nutrients inefficiently and some of them may end up polluting water or air. For every acre on which they applied manure, confined animal feeding operations added 76 pounds of excess phosphorus versus only 4 pounds for small farms. CAFOs generated up to 832 pounds of excess nitrogen compared to a maximum of 125 pounds per acre for small farms (Loecke and Jackson 2001). Farms that rely on tight nutrient cycling for soil fertility instead of fertilizer will have a major incentive to keep those nutrients out of the streams. CAFOs often have incentives to get rid of manure as cheaply and quickly as possible.

Restoration of Prairie Water Quality and Hydrology

Recent measures of water quality show that Iowa's lakes and rivers have some of the highest levels of nitrates and dissolved phosphorus in the world. Most Midwestern streams are impaired by excess sediment, which disrupts the lifecycle of fish such as the federally endangered Topeka shiner. Restoration of long crop rotations will facilitate serious reductions in fertilizer and pesticide use both on a per-acre basis (corn after clover or alfalfa requires far less fertilizer than corn after soybeans) and on the number of acres treated (fewer acres of corn and soybeans). This would begin to address our water-quality problems at their source.

Complete hydrologic restoration will remain a distant goal, however. Cropped fields (many ditched or drained with subsurface tiles) will always hold less water than prairie. Water is forced to leave the landscape faster than it did on the prairie and therefore will require streams capable of handling that speed. Wetland restoration to slow runoff is an option, but to fully contain these waters would mean more wetlands than before farming, and not less. One need only look at the hydric soil distribution on county soils maps to see that in some regions perhaps 80 percent of the currently cropped land was originally wetland. According to early ecologist Thomas MacBride (1895),

> The prairies were wet, and in all low places stayed wet. Very rarely did the surplus water pass off by anything like a ditch as now, but every valley was a bog, utterly impassable to man or beast. The waters did not seem to run at all, but gradually evaporated or sank to lower and lower strata. . . . Over the oozy sloughs the sedges waved head-high, and into their treacherous depths horses, oxen or even men ventured at peril of their lives.

Restoration of the wetter, more permanent wetlands and those near frequently flooding rivers will help, but it is unlikely that the vast ephemeral wetlands of the prairie states will be released from agricultural production any time soon. Therefore, it is important that we find appropriate agricultural uses for these areas. If they are not planted in row crops, they can continue to receive water, temporarily slow it down, and filter it somewhat as a natural ecosystem would. Lowlands might accommodate intensively managed rotational grazing or annual haying. Some of the best prairie remnants in Iowa were wetlands, managed for wild hay into the 1950s. Farmers could only justify converting broad swaths of land along streams to hay if public policy rewarded them for protecting public water resources (see chapter 18). Current incentives to create "buffer strips" around streams are a step in this direction.

Protection and Restoration of Margins, Edges, and Fragments

Crop rotations may restore some heterogeneity to the landscape but will not guarantee that native biodiversity is preserved. Several other actions will be necessary to protect and restore the dwindling natural areas:

- *Restoration of marginal land.* Some lands are truly too steep or wet to grow row crops at all, and yet by dint of extraordinary engineering and wishful thinking they are planted with row crops anyway, even if they lose money. The large effort to enroll land owners in conservation reserve programs on highly erodible land over the last fifteen years has been helpful in this regard. There is an exciting new program administered by the U.S. Fish and Wildlife Service to enroll 125,00 acres of small (8- to 20-acre) farmed wetlands into conservation reserve in Iowa's Prairie Pothole landform. It is hoped that once the ten-year sign-up period is over, farmers will opt to keep the wetlands in place. It might be better for conservation and more cost effective if we were able to simply purchase the conservation easement to take them out of production permanently.

• *Protection of remaining habitat remnants on farms.* Despite the loss of 99.6 to 99.9 percent of tallgrass prairie habitat (Samson and Knopf 1994), small remnants of native prairie persist on private land. The U.S. Fish and Wildlife Service has proposed the creation of a 77,000-acre Northern Tallgrass Prairie Habitat Preservation Area over twenty-five years for western Minnesota and northwestern Iowa (USFWS 1998). The Service will work with individuals, groups, and government entities to permanently preserve remnant tracts of prairie. Landowners may want to sell the land outright, while others may be willing to sell a conservation easement or simply modify their management practices to better protect native species.

This will require educational outreach to rural residents to make them aware of the natural heritage on the back forty. In informal visits with landowners about protecting what is left of Iowa's natural communities, I often find indifference or suspicion. Many think a prairie is a bunch of weeds. This is partially the fault of the university and amateur botanists and photographers who cruise the prairies in the spring to enjoy their favorite flowers. Many of these prairies are well-kept secrets known only to the cognoscenti. We have failed to engage with people in rural communities, perhaps because they are outside of our accustomed social circles. We have also worried, rightly, that if we make these prairies better known, local residents will only dig up the flowers for their own gardens. Still, the cost of keeping secrets is high. In just eight years in northeast Iowa, I have learned of at least five prairie remnants that were plowed up when the property changed hands. Within the Northern Tallgrass Prairie Habitat Preservation Area, there will be a conservation buyer (the U.S. Fish and Wildlife Service) ready to give landowners another option, but elsewhere there is essentially no system in place to help landowners recognize and protect ecologically significant lands.

• *Restoration of roadsides and other publicly owned lands.* An ambitious state and federally funded program is now underway to make native seeds available for roadside plantings. Education programs help maintenance workers learn to recognize—and refrain from spraying—native roadside wildflowers. In Iowa at least 20,000 acres of rural roadsides have been seeded with native prairie species in the last fourteen years, and the number of participating counties increases every year (Houseal and Smith 2000). Bike trails, railroad tracks, county, state, and federal lands, and even urban areas, are all gaining restored prairie. These areas are very tiny but relatively important as potential corridors. Their importance for public education is incalculable.

- *Enhancement of remaining pasture and hay lands with native plants.* By using rotational grazing techniques, it may be possible for farmers in the former prairie states to establish and maintain some subset of the native prairie vegetation—along with attendant insects and other animals—and still derive an income from the land (Jackson 1999). After all, native prairie is used for cattle grazing in large parts of the Great Plains. Grazed prairie pasture falls short of full-fledged prairie, but it may help provide the landscape connectivity needed for dispersing and migrating animals throughout the Upper Midwest.
- *Slow the spread of exotic species.* Although we worry about creating more connections among habitats for native species, there is another category of plants and animals that travel all too well in the modern landscape. Purple loosestrife, garlic mustard, leafy spurge, the zebra mussel, kudzu, and many other exotic species hitchhike on people, their livestock, and their cars, boats, planes, and trains, invading wherever they land. Exotic species invasion is a global phenomenon that has cost billions of dollars and is the number one cause of endangerment in the United States for species classified as threatened or endangered by the U.S. Fish and Wildlife Service (Vitousek et al. 1996, Wilcove et al. 1998). Somehow, our modern landscape crisscrossed by road, rail, and barge, is going to have to become *less* easy for these species to traverse if we wish to avoid the substantial costs of invasive species.

Conclusions

It would be foolish to claim that crop rotations as practiced in the Midwest before the 1950s solved every problem of agriculture. Under traditional crop rotations before World War II, the level of nitrates in the Des Moines River was already high (Keeney and DeLuca 1992). Plowing and as many as seven cultivations each year for weed control created high rates of soil erosion on the land devoted to corn. Farmers were already on the path of industrialization, and they no more viewed their land as an "ecosystem" than most farmers do today. Leopold (1966) describes Illinois in the 1940s:

> Everything on this farm spells money in the bank. . . . The old oaks in the woodlot are without issue. There are no hedges, brush patches, fencerow or other signs of shiftless husbandry. The fences stand on narrow ribbons of sod; whoever plowed that close to barbed wires must have been saying, "Waste not, want not."

What the farmers of the early twentieth century had right was more by virtue of constraint than restraint. Farm productivity depended by necessity upon ecosystem services combined with clever manipulation. Nevertheless, the favorable comparisons between prairie ecosystems and early Midwestern agroecosystems suggest that we may be able to design agroecosystems that do an even better—and more purposeful—job at mimicking natural ecosystems while growing our food.

One characteristic of ecological restoration is that if key ecosystem processes are established, such as the appropriate disturbance regime, many native plants and animals will again be able to find the area and thrive there. Wendell Berry (1981) called it "solving for pattern." Restoration of long crop rotations linked to livestock production in the Upper Midwest could restore several elements of the prairie ecosystem: year-round physical protection of the land by perennial plants; a diversity of plants whose life cycles take advantage of different seasons of the year; a food web that includes broadly grazing ruminant herbivores; a landscape dominated by herbaceous perennial grasses and forbs (in the form of hay and pasture); nitrogen from manure and legumes, and nitrogen delivery to plants coming principally from mineralization of organic matter synchronized with crop nutrient uptake. Prairie-like landscape connectivity, hydrology, nutrient cycling, soil structure and bird life—these are all elements of prairie worth restoring.

References

Berry, W. 1981. *The Gift of Good Land: Further Essays Cultural and Agricultural,* pp. 134–145, North Point Press, San Francisco.

Best, L. B., K. E. Freemark, J . J. Dinsmore, and M. Camp. 1995. "A Review and Synthesis of Habitat Use by Breeding Birds in Agricultural Landscapes of Iowa." *American Midland Naturalist* 134:1–29.

Bulkley, R. V. 1975. *A Study of the Effects of Stream Channelization and Bank Stabilization on Warm Water Sport Fish in Iowa, Subproject No. 1, Inventory of Major Stream Alterations in Iowa.* Completion Report, U.S. Fish and Wildlife Service, Contract No. 14-16-0008-745.

Dinsmore, J. J. 1994. *A Country So Full of Game*. University of Iowa Press, Iowa City.

Hayden, A. 1947. "The Value of Roadside and Small Tracts of Prairie in Iowa as Preserves." *Proceedings of the Iowa Academy of Science* 54:27–31.

Houseal, G. and D. Smith. 2000. "Source-identified Seed: The Iowa Roadside Experience." *Ecological Restoration* 18: 173–183.

Hurley O'Hara, J. 1999. "Nesting Bird Use in Native Prairies, Savannas, and Pasture Agroecosystems." Master's thesis, University of Northern Iowa, Department of Biology, Cedar Falls.

Iowa Agricultural Statistics Service. 1994–2000. *Iowa Agricultural Statistics/ compiled by Iowa Agricultural Statistics*. U.S. Department of Agriculture, National Agricultural Statistics Service and Iowa State University Extension to Agriculture, Washington, D.C.

Iowa Agricultural Statistics. 1986–1990. *Iowa Agricultural Statistics/ compiled and issued by Iowa Agricultural Statistics*. Iowa Crop and Livestock Reporting Service, Des Moines.

Iowa Department of Agriculture. 1900–1985. *Annual Iowa Year Book of Agriculture*. Iowa Department of Agriculture, Des Moines.

Iowa Executive Council. 1854–1925. *Census of Iowa*. Iowa General Assembly, Des Moines.

Iowa State Agricultural Society. 1865–1899. *Annual Report of the Board of Directors of the Iowa Agricultural Society*. Iowa State Agricultural Society, Des Moines.

Jackson, L. L. 1998. "Agricultural Industrialization and the Loss of Biodiversity." Pp. 66–76 in *Protection of Global Biodiversity: Converging Strategies*, edited by L. D. Guruswamy and J. A. McNeely. Duke University Press, Durham, N.C.

——. 1999. "Establishing Tallgrass Prairie Species on a Rotationally Grazed Permanent Pasture in the Upper Midwest: Remnant Plant Assessment and Seeding and Grazing Regimes." *Restoration Ecology* 7:127–138.

Jackson, L. L., D. Keeney, and E. M. Gilbert. 2000. "Swine Manure Management Plans in North-Central Iowa: Nutrient Loading and Policy Implications." *Journal of Soil and Water Conservation* 55:205–212.

Jackson, L. S., C. A. Thompson, and J. J. Dinsmore. 1996. *The Iowa Breeding Bird Atlas*. University of Iowa Press, Iowa City.

Jackson, W. 1980. *New Roots for Agriculture*. University of Nebraska Press, Lincoln.

——. 1987. *Altars of Unhewn Stone: Science and the Earth*. North Point Press, San Francisco.

Keeney, D. R., and T. H. DeLuca. 1992. "Des Moines River Nitrate in Relation to Watershed Practices: 1945 Versus 1980s." *Journal of Environmental Quality* 22:267–272.

Lefroy, E. C., R. J. Hobbs, M. H. O'Connor, and J. S. Pate. 1999. *Agriculture as a Mimic of Natural Ecosystems*. Kluwer Academic Publishers, Dordrecht.

Loecke, T., and L. L. Jackson. 2001. "Uncoupling Nutrient Cycles in Midwestern Agroecosystems: N and P Flows on Diversified Farms Versus Confined Livestock Operations." Abstract, Annual Meeting of the Ecological Society of America, Madison, Wis., August 5–10, 2001.

Logsdon, S. D., J. K. Radke, and D. L. Karlen. 1993. "Comparison of Alternative Farming Systems. I. Infiltration Techniques." *American Journal of Alternative Agriculture* 8:15–20.

MacBride, T. 1895. "Landscapes of Early Iowa." *The Palimpsest* 11:283–293.
National Agricultural Statistics Service (NASS). 2000. *2000 Agricultural Statistics*. U.S. Department of Agriculture, U.S. Government Printing Office, Washington, D.C.
Pammel, L. H., J. B. Weems, and F. Lamson-Scribner. 1901. *Pastures and Meadows of Iowa*. Iowa Agricultural College Experiment Station, Ames.
Quick, H. 1925. *One Man's Life, an Autobiography*. Bobbs-Merrill, Indianapolis.
Randall, G. W., D. R. Huggins, M. P. Russelle, D. J. Fuchs, W. W. Nelson, and J. L. Anderson. 1997. "Nitrate Losses through Subsurface Tile Drainage in Conservation Reserve Program, Alfalfa, and Row Crop Systems." *Journal of Environmental Quality* 26:1240–1247.
Randall, G. W. 2000. "Row Crops Can Have Thirty to Fifty Times Higher Nitrate Losses Than Perennials." *News and Information*, November 11. University of Minnesota Extension Service, St. Paul.
Samson, F. B., and F. L. Knopf. 1994. "Prairie Conservation in North America." *BioScience* 44:418–421.
Shimek, B. 1917. "The Persistence of the Prairie." *University of Iowa Studies in Natural History* 11:3–24.
Smith, D. D. 1992. "Tallgrass Prairie Settlement: Prelude to the Demise of the Tallgrass Ecosystem. Pp. 195–199 in *Proceedings of the Twelfth North American Prairie Conference*, edited by D. D. Smith and C. A. Jacobs. University of Northern Iowa, Cedar Falls.
Soule, J. D., and J. K. Piper. 1992. *Farming in Nature's Image: An Ecological Approach to Agriculture*. Island Press, Washington, D.C.
U.S. Fish and Wildlife Service (USFWS). 1998. *Northern Tallgrass Prairie Habitat Preservation Area Final Environmental Impact Statement*. U.S. Fish and Wildlife Service, Fort Snelling, Minn.
Vitousek, P. M, J. D. Aqber, R. W. Howarth, G. E. Likens, P. A. Matson, D. W. Schindler, W. H. Schlesinger, and D. G. Tilman. 1997. "Human Alteration of the Global Nitrogen Cycle: Sources and Consequences." *Ecological Applications* 7:737–750.
Vitousek, P. M., C. M. D'Antonio, L. L. Loope, and R. Westbrooks. 1996. "Biological Invasions as Global Environmental Change." *American Scientist* 84:468–478.
Wilcove, D. S., D. Rothstein, J. Dubow, A. Phillips, and E. Losos. 1998. "Quantifying Threats to Imperiled Species in the United States." *Bioscience* 48:607–616.

Chapter 11

Sustaining Production with Biodiversity

Nicholas R. Jordan

At the heart of farming is a two-fold challenge. To succeed in farming, many now argue, one must produce profitably while also being a "good neighbor": a contributor to the well-being of surrounding communities and landscapes. To meet this challenge, farms must function as *both productive* and *protective* ecosystems. Productive ecosystems produce a substantial quantity of biomass on an ongoing and cost-effective basis. Protective ecosystems tend to be self-contained, and resources such as water, soil, or nutrients are tightly held and rarely disgorged to surrounding lands. Alas, production and protection are not easily reconciled. Farmers must find land for protective features, such as plantings around watercourses. They must also provide large seasonal "pulses" of nutrients needed for abundant production while preventing pollution from loss of these nutrients from the farm. It is easy to see protection as the antithesis of production. Yet, the contradiction resolves in a synthesis: ecosystems that contain enough of the right sorts of biodiversity will be, in good measure, *both* productive and protective (Altieri 1999).

Therefore, one fundamental goal of agroecological restoration must be development of farms that contain such biodiversity. Several other chapters in this book will propose strategies for restoration and for the new working relationships needed to implement such strategies. First, however, we must consider how it is that biodiverse agroecosystems can be both productive and protective by focusing on two fundamental issues in farming: soil quality and pest control. Farmers must have a soil that has sufficient fertility and workability, and pests must be controlled. Modern agriculture has discovered how to meet these needs—in the short term—by mechanization, irrigation, and large inputs of synthetic fertilizers and pesticides (Matson et al. 1997). However, it is very clear that, without biodiversity, these needs cannot be met in the long run.

Biodiversity and Soil Quality

Soils of high quality supply sufficient water and nutrients for bountiful crops while resisting degradation due to wind, water, and disturbance from farm operations (Carter et al. 1997). The fundamental source of soil quality is the work of living things. At heart, the biological creation of high-quality soil results from a partnership between plants and the inhabitants of the soil around their roots (the soil biota); this partnership has been called "biotic regulation" of soil (Swift and Anderson 1993, Perry 1995).

In this partnership, the plant provides resources to the soil biota in two forms (Swift and Anderson 1993): *litter* (dead material), and *root exudates* (nutritious substances released to surrounding soil or transferred directly to soil biota that are physically connected to the plant root). Plants also cultivate soil biota by controlling soil temperature, moisture, and other environmental qualities. In return, the soil biota, collectively, provide many benefits to the plant, thereby supporting its growth. This growth allows the plant, in turn, to provide an increasing flow of resources to the soil biota. In this way, plants and soil biota are linked together in a relationship of mutual support (although interlopers, such as parasitic organisms that cause soil-borne plant diseases, can enter the scene). Benefits received by plants from the work of soil biota include capture and storage of essential nutrients (nitrogen, phosphorus, etc.) and release of nutrients when most needed for plant growth, as well as protection from diseases and stresses such as drought (Swift and Anderson 1993).

Plants also benefit from activities of soil biota that develop soil quality by effects on the physical structure of soil. For example, certain soil biota—the mycorrhizal fungi—act to "glue" smaller soil particles into larger "crumbs," increasing availability of water and oxygen to roots (and to soil biota) (Hooker and Black 1995). Also, such soils are more able to capture and store rainfall for future use and to protect organic matter (the store of essential plant nutrients in the soil) from decomposer organisms that compete with plants for these nutrients (Miller and Jastrow 1990). These effects of the plant-soil partnership make it clear that biotic regulation is a powerful source of soils whose quality makes a farm both productive and protective. Moreover, biotic regulation has the great virtue of being a "self-starter." As I will argue below, it appears that biotic regulation will go to work producing soil quality provided only that a certain diversity of plants and soil biota is present (Perry et al. 1989).

Soils whose quality is maintained by human inputs are harder pressed to meet this standard (Matson et al. 1997). First, consider soil fertility. Ideally, fertility must be provided cost effectively while limiting unwelcome leakage of nutrients into neighboring ecosystems. Synthetic fertilizers—the prevalent fertility source in high-yield farming—are unsatisfactory in both regards. Large proportions (often greater than 50 percent) of applied synthetic fertilizer are lost from farms, often causing serious pollution of surrounding lands and waters (Andow and Davis 1989). These lost nutrients produce nothing for the farmer and often mean that fertilizer use is not cost effective. In contrast, when biotic regulation of soil is at work, nutrients are mostly held in "resistant" forms that are not easily lost from the soil (Swift 1997). At times when plants have a large need for nutrients they will stimulate the activity of the soil biota causing nutrients to be rapidly converted from their resistant forms. Thus, biotic regulation of soil meets the need for fertility in a much more cost effective and "neighborly" manner than through use of synthetic fertilizers alone. As Laura Jackson has discussed in chapter 10, this is a restoration of plant-soil biota relationships found in the prairie and other natural ecosystems.

Farming without biotic regulation of soil has even more damaging effects on other aspects of soil quality. The tilth of a soil—its texture, and related properties such as water-holding capacity—can degrade rapidly without biotic regulation. The "glues" mentioned above that aggregate soil into larger "crumbs," are critical to soil tilth. Glues are produced by certain soil biota that are dependent on energy subsidies from plants. Tilth depends on ongoing production of these glue substances, which are not durable. Sadly, synthetic fertilizer use destroys tilth, because fertilizers cause physiological changes in crop plants that induce them to withhold subsidies from their soil biota (Hooker and Black 1995). To understand how this great misfortune occurs, recall that the biota can supply nutrients to their plant partners in return for an energy subsidy. However, if nutrients from synthetic fertilizer are readily at hand, the biota are not needed to supply nutrients, and crop plants, probably as a result of modern plant breeding, respond by cutting off the energy subsidy and thus the production of tilth-building glues.

Loss of tilth robs a farm of both productive and protective qualities. As tilth declines, a soil loses ability to retain rainfall and quickly becomes waterlogged—a condition in which the soil lacks oxygen because all of the empty space in the soil is filled with water. Again, both farmer and neighbors suffer. For the farmer, serious problems arise in both wet and dry years. In wet years, waterlogged soils kill plant roots and cause rapid loss of nutrients from soil. In dry years, the limited rainfall is likely to run off rather than to be retained by the soil so that water quickly becomes limiting to plant growth after a rain. For neighbors, rapid runoff of rainfall produces highly variable conditions in neighboring waterways: floods are followed by dry streambeds. Rapid runoff makes a soil much more vulnerable to erosion of soil particles, further damaging the farm and its neighboring ecosystems.

Biodiversity and Crop Protection

Globally, it is estimated that some 30 to 40 percent of farm production is currently lost to pests before harvest (Tivy 1990). Pest control is the leading concern of mainstream agricultural science. Remarkably, the massive pesticide use typical of high-yield farming does not seem to have reduced losses much and it is widely agreed that pesticides cannot, by themselves, provide durably effective and affordable protection of crops from pests (Pimentel et al. 1991). There is also wide agreement that biological diversity must be brought to bear on the pest problem, along with non-pesticidal methods, such as cultivation, to kill weeds (Altieri 1999).

The most promising general approach to improving pest control rests squarely on biodiversity. It is called "conservation biocontrol," and it aims to conserve sufficient biological diversity on farms and in farmed landscapes to keep pest populations at or below acceptable levels (Letourneau 1998; see also chapter 9). The necessary biodiversity consists of the pests themselves, their "regulators" (e.g., predators and parasites) that consume pests, alternate food sources for the regulators for times when the pest is not available or abundant, and plants that provide habitat for all parties. All of these actors (at least) seem to be necessary for ongoing regulation of pests. Although the pests themselves may be abundant, the other actors are often in short supply on the farm and in farmed landscapes, often because most of the land area is used directly for production (i.e., to grow crops), leaving little habitat for necessary biodiversity. This conflict between land and habitat complicates the practice of conservation biocontrol. The craft and science of creating, improving, and maintaining sufficient habitat to enable conservation biocontrol without excessive cost is now being developed by farmers and scientists worldwide. For example, it is clear that vegetation that is "out of the way" in time or space—such as cover crops that are planted after harvest, or plantings along field edges—can maintain pest regulators by providing alternate sources of food and shelter during seasons when neither crop nor pest is present (Bugg 1992). Such strategies of conservation biological control have great potential by helping farms more closely approach the productive/protective ideal and reducing their reliance on pesticides.

In sharp contrast, pesticide-based pest control has many well-known problems (Andow and Davis 1989, Matson et al. 1997). First, use of pesticides is often harmful to neighboring ecosystems and people. Second, pesticide control cannot sustain cost-effective production. Successful control of one pest often creates increased opportunity for another, for example, to consume newly available resources or to escape control by predators and parasites that were unintentionally killed by the pesticide. These effects of pesticides create the notorious pesticide "treadmill," by which farmers relying on pesticides often find themselves using ever-increasing quantities to achieve acceptable control. Finally, pesticide control threatens production by making farms more vulnerable to disruptions. For example, pesticide control often only works when pesticides can be applied during a certain period of pest vulnerability. If weather conditions or labor demands prevent application within this period, there may be no control option whatsoever, leading to crop failure.

By contrast, conservation biocontrol has very few of these problematic effects. Habitat created for conservation is likely to have a net positive effect on surrounding ecosystems. For example, streamside plantings can keep soil and nutrients from escaping into streams. Conservation biocontrol is likely to be less vulnerable to disruption, because it aims to maintain a diversity of pest regulators. In theory, this provides a diversity of controlling influences on pest populations, limiting the likelihood of outbreaks of any of the set of pests present and compensating, through redundancy, for situations (e.g., certain weather events) that may reduce the effectiveness of any one regulator (Altieri 1999). They are also likely to be more cost effective: typically, resource inputs to create and maintain conservation habitat will produce more pest control per unit input than pesticide use. Moreover, resource inputs to create habitat will produce other benefits, such as water-quality protection and enhancement of other biodiversity, including wildlife.

Ecological Partnerships: A Touchstone for Restoration

There is an important lesson to be learned from studying the links between biodiversity, soil quality, and crop protection. Biodiversity provides production and protection not through the work of single species, but rather by ensembles of species that work in *ecological partnership* (Crossley et al. 1984). For example, such a partnership—between plants and their associated soil biota—is at work in the biological maintenance of soil quality. Given a certain amount of care through appropriate farming practices such as use of cover crops, the partnership is self-maintaining and provides extremely valuable services on an ongoing basis. Increasingly, ecologists are finding that such partnerships provide the *integrity* of ecosystems—that is, the glue that holds them together in the face of natural vagaries great and small (Perry 1995). For example, partnerships appear to enable ecosystems to retain their stocks of limiting resources such as water and nutrients, and to recover from destructive events such as floods, droughts, fires, and storms.

Ecological partnerships appear to be at work on many scales. For example, it is now clear that plant leaves create a habitat—known as the phyllosphere—in which complex microbial communities exist, subsisting on resources that occur on plant leaves and physically protected by microscopic features of the leaf surface (Kinkel 1997). Recent research has shown that if certain microbial communities are established on leaf surfaces, then these communities are capable of excluding disease-causing microbes. This insight teaches that biocontrol of plant disease can be achieved by providing conditions that foster certain microbial communities on the leaf surface. Thus, the plants and their phyllosphere biota form an ecological partnership that limits losses to plant disease.

The range of what we understand as ecological partnerships has been expanded by recent discoveries. Conventionally, relations between living things have been divided into situations where both parties benefit—*mutualisms*—and situations where parties have conflicting interests—*antagonisms*, such as competitive, predatory, or parasitic relationships. However, even seemingly antagonistic relationships can be mutually beneficial when these relationships are viewed more broadly, as parts of a larger partnership. Recently, it has been shown that plants can actually benefit from animal grazing provided that the animals return nutrients to soil around the same plants that they feed upon. Grazing will benefit the plant if it makes these nutrients available to the plant more quickly and efficiently than otherwise (De Mazancourt and Loreau 2000).

Despite their fundamental importance, our predominant images of nature do not feature ecological partnerships. Therefore, considering the place of partnerships in farming may require a significant shift in world view. Instead of thinking of the farm as populated with individual organisms—crops, livestock, wildlife, and weeds—it is much more useful to see these beings as members of partnerships (Swift et al. 1996). Moreover, there is good reason to think that farmers can develop ecological partnerships that are much more beneficial than those that are typical of current high-input farms. Through artful husbandry, farmers may be able to create conditions in which evolutionary and ecological processes work to weave species into the mutually beneficial webs and cycles that make for effective partnerships (Wilkinson 1998).

Moreover, if partnerships are important, then it follows that their development should be at the heart of agroecological restoration that aims to fully engage biodiversity in farming. Accordingly, the development of partnerships through artful husbandry offers farmers a powerful strategy for agroecological restoration. For example, it is becoming clear that relations between crops and their associated soil biota are highly contingent on husbandry. If farmers maintain high levels of plant diversity on their farm and minimize soil disturbance and synthetic fertilizer use, then the abundance of plant-beneficial soil biota appears to increase (Hooker and Black 1995). Conversely, as noted above, the use of large quantities of synthetic fertilizers appears to fundamentally change the plant–soil-biota relationship. A shift occurs, from a situation of mutual benefit to one in which some of the soil biota apparently become damaging parasites of crop roots and probably also contribute little to the maintenance of soil quality (Johnson 1993). Thus, it is clear that the plant–soil-biota partnership will not automatically benefit a farm—the right type of husbandry is required to develop and maintain a beneficial partnership and thus to restore a crucial part of the integrity and functionality of the agroecosystem.

In the next section, I will argue the value of three broad approaches to development of effective ecological partnerships on farms. First, I will suggest that ecological partnerships can be *improved by building them in place* by certain processes of genetic evolution. Second, I will argue that effective partnerships, once formed, are constantly threatened with disruption because of the natural variability of ecosystems. Therefore, farms must be *designed to protect partnerships* against this threat. Finally, I will contend that *new working relationships* among farmers, scientists, and other agriculturalists are necessary for development of the knowledge of biodiversity and farming needed for these and other avenues of agroecological restoration.

Development of Ecological Partnerships

The close working relationships in effective partnerships are now believed by biologists to require that the partners adapt to partnership via genetic evolution (Herre et al. 1999). Recently, we have learned that the necessary adaptation can occur rapidly, over a matter of years or decades; in other words, quickly enough to be part of practical agroecological restoration (Thompson 1998). However, evolution of effective partnerships has a puzzling, paradoxical aspect: partnership usually involves some form of self-limiting behavior by the partners. For example, partners in a mutualism often exchange goods or services, limiting the "profit" they realize in the transaction so that their partner benefits. The question is how such self-limiting behavior could evolve by any process of "survival of the fittest." It is becoming clear that this evolution could occur in several different ways

First, some species might directly benefit by cooperative behaviors in a partnership despite some measure of self-limiting behavior. For example, mycorrhizal fungi are believed to withdraw nutrients from dying crop plants and to transport them to germinating seedlings of a cover crop. An individual fungus behaving thusly could directly benefit by such a prudent "investment" of nutrients, because the cover crop will mature to support the future growth of the fungi (Wilkinson 1998). Happily, such fungal behavior would also reduce nutrient losses from the farm.

Evolution might also work by another pathway, sometimes called group or community selection, in which groups of organisms cooperate in an ecological partnership in a way that, unlike "non-partners" (individuals of the same species that do not cooperate as fully), increases the ability of all group members to survive and reproduce (Wilson 1997). In this scenario, the partners will increase in frequency relative to non-partners, and therefore successful partnerships will spread. In the past, most ecologists have regarded community selection as a weak force. The advantages gained by cooperation were thought to be too small—and the advantages gained by "cheaters" who avoided self-limiting behavior too large—to allow partnerships to spread very much. However, farms and other ecosystems that are looked after by people are different in a fundamental way. Through appropriate husbandry, people can provide conditions that provide a strong advantage to effective partnership, and they can maintain these conditions over time.

How might this process of community selection work in practice? For example, imagine that a crop is growing in a field with low levels of plant-available phosphorus in the soil. A number of crop species (e.g., wheat, corn, and soybean) are known to contain some genetic variants that are willing cooperators with mycorrhizal fungi, and other genetic variants that spurn associations with the fungi. Imagine such a range of crop genotypes growing within the field. Imagine also a range of fungal genotypes varying from genetic types that provide little phosphorous in return for their carbon subsidies (poor "mutualists") to types that provide richer rewards. A range of plant-fungus combinations will form—each making up a plant–soil-biota community. However, in this situation, only the good host/mutualist combination will do well. All combinations with a poor plant host will do poorly because the plants will lack essential phosphorus. All combinations with a poor mutualist will also lack phosphorus, or pay too dearly for it. The good host/mutualist combination will grow vigorously, forming a large root system and thus allowing the fungus (and the food web of soil biota associated with it) to grow large and reproduce heavily.

Finally, imagine that a farmer is watching this field, intent on saving seed from the "best" plants for the next generation. In fact, the best plants are actually the best plant-fungus (or, more likely, the best plant-fungus-food web) combination. As long as the "best" seeds are planted back into this field, the good host/mutualist combination will form again for another cycle of increase. Eventually the good host/mutualist combination will prevail throughout the field because of this process of community selection. As a result, we will have an evolved plant-soil partnership that improves the productive and protective qualities of the farm.

This example shows how community selection could improve the performance of agroecosystems by strengthening an ecological partnership. Beyond the fundamentally important partnership between plants and soil organisms there are many other relationships that might be significantly improved by a community selection process. These include relationships between plants and the microorganisms that occur on their surfaces or within their above-ground parts, and relationships within groups of social animals. For example, chicken breeders have recently experimented with selection of flocks of hens on the basis of the egg production of the whole flock. The result was more productive flocks, apparently because the resulting hens were less antagonistic to each other (Goodnight and Stevens 1997). Other relationships that might be improved include plants with their pollinators, pests with their natural enemies, parasites with their hosts, and plants with their herbivores.

The question becomes: how can farmers act to set community selection in motion so as to improve ecological partnerships? What husbandry could have this effect? We can tentatively state that the very important mutualism between crops and mycorrhizal fungi can be maintained by avoiding practices that harm the fungi or cause the plants to harm them. As noted above, harmful practices include large inputs of synthetic fertilizers, high levels of soil disturbance, bare-soil fallow periods, and frequent rotations to non-host crops (Hooker and Black 1995). Similar guidelines are needed for other ecological partnerships; their identification and improvement should be a major focus of agricultural science.

Maintenance of Ecological Partnerships

If we accept that ecological partnerships are important to farming, then we must also consider another issue: how are these to be sustained in the face of threats to their integrity? Threats arise from disturbances that are part and parcel of farming (e.g., plowing of soil or harvest of plants) and others that are extrinsic (e.g., droughts, floods, biotic invasions, or social and economic change). Such factors may directly harm one or more members of a partnership and may disrupt the ongoing interactions that sustain the relationship. Studies of the maintenance of partnerships in other ecosystems suggest that valued partnerships on farms can be protected by a second strategy of agroecological restoration: creating certain *patterns* of biodiversity in time and in space. Certain patterns seem crucial to the protection of ecological partnerships in forest ecosystems; these patterns may be effective on farms as well.

One important form of pattern is the *biological legacy*. In this case, an ecosystem is patterned so that key portions of ecological partnerships are safely sheltered during a disturbance, allowing subsequent regeneration (Perry

1995). The sheltered remnants constitute a "legacy" from the pre-disturbance ecosystem. This patterning is necessary when one or more partners cannot withstand the disturbance except when sheltered. An example of such a pattern is an abundant scattering of large fallen tree trunks—termed large woody debris by foresters—in certain forests of the Pacific Northwest. Regeneration of these forests after fire requires certain tree species and their mutualists among the soil biota. The trees can survive a moderate fire readily, as buried seeds. However, the soil biota can survive only when sheltered from harsh post-fire soil conditions. In these forests, only large fire-resistant fallen logs provide such shelter. After fire, tree seedlings germinating around the periphery of these logs are colonized by their mutualists and can regenerate the forest while seedlings that are beyond the reach of the legacy cannot. Without an adequate patterning—a sufficient amount of large dead wood, scattered broadly—the forest will not recover from fire.

Another common pattern is termed a *cooperative guild* (Perry 1995). Here, a group of organisms is spread widely enough in time and space to collectively maintain the viability of a shared partner species. For example, consider a group of plant species, each dependent on a common pollinator. If the species within the group flower over the full period when the pollinator is active, then the group—as a whole—will maintain the pollinator. Cooperative guilds apparently serve to protect partnerships between guild members and a common mutualist against a range of disturbances. This broad protection will occur if the guild includes at least one species that is tolerant of each of the disturbances and can play the role of sustaining the partnership in the event of the disturbance to which it is adapted.

Biotic legacies and cooperative guilds are archetypes of patterned biodiversity that sustains ecological partnerships. Many current agroecosystems, for instance, most high-input cropping systems in developed countries, lack such patterning and consequently are probably unable to provide protection to partnerships. Many farmers who work in these situations are experimenting in search of effective and economically affordable patterning. For example, the emerging craft of farmscaping aims to create such patterns.

Farmscaping is being developed by growers of vegetables and fruit crops in the challenging environment of California, where climate and intensive agriculture have resulted in severe plant pest problems. It involves the deployment of patterns of varied plantings across the terrain and the temporal calendar of a farm with the intention of supporting and stabilizing populations of predators and parasites, which in turn act to regulate populations of various pests of crop production. These plantings, by virtue of their spatial or temporal location, serve to maintain populations of pollinators, predators, and parasites by providing shelter or food at times when they are critically needed by these populations and are not otherwise available on the farm or in the surrounding landscape. Quite possibly, these plantings also harbor other organisms (e.g., diseases and parasites) that serve to maintain the populations of predators and parasites within a certain range. For example, farmscapers are very aware of the value of strips of vegetation that grow along the edges of fields. If these are actively growing during times of year when adjacent fields are bare, then these strips can support pollinators and other organisms that disperse into adjacent fields containing crops.

New Working Relations Among Agriculturalists

Agroecological restoration of biodiverse farms—via the strategies outlined above or by other approaches—will require development of new social relationships that can produce the knowledge needed for restoration. These new relationships are needed because of several distinctive features of this sort of farming.

First, such farming is likely to be highly site specific. That is to say, how biodiverse farms are managed will likely depend on the particular combination of land, available biodiversity, and commercial enterprise present on a farm. Each of these three factors must be taken into account in the restoration process. In essence, restoration is a challenge of design and operation that must be solved farm by farm. We must recognize that individual farmers have essential roles as designer-operators because only they can develop a sufficient knowledge of the interplay on their farms between land, organisms, and enterprise. Their ability to function as well-informed, fast-learning designer-operators is critical to successful restoration and is a daunting challenge that can be likened to redesigning an airplane while flying it.

The question then becomes, how can we develop the ability of interested farmers to be effective designer-operators of biodiverse farms? I believe that we must intentionally search for ways of facilitating learning processes that will develop these abilities. Learning appears to be necessary on several scales. First, individual farmers must develop skills that enable them to be so-called "adaptive managers." In essence, adaptive management of farms (or any other complex situation) involves making an *inquiry cycle* integral to management. In the inquiry cycle, problem situations are identified, a hypothesis (or several) is formed regarding what actions might improve the situation, actions are taken guided by this hypothesis, and results are observed. Based on these results, the hypothesis is modified and another cycle of action and monitoring begins. In chapter 14, Rhonda Janke describes this process in the context of whole farm planning.

Such individualized learning by farmers is probably critical to agroecological restoration. However, it seems likely that individual learning must also be complemented by learning at larger scales in which groups of farmers learn cooperatively. For example, a set of farmers can cooperate to screen a range of cover-crop species to reduce a large group of potentially useful species to a smaller set of more promising species. Cooperative learning is also needed in dealing with problems that occur at scales larger than individual farms in which coordinated actions by many farmers are needed to improve the situation. Farmers are experimenting with cooperative learning in many places globally; examples include Central American farmers (Holt-Gimenez 1999) and European farmers involved in intensive horticultural production (Somers 1998). Brian DeVore discusses cooperative learning strategies of farmers in chapter 8.

In addition to the roles of farmers in individual and cooperative learning, there are important roles for other agricultural workers. For example, many farmers are convinced that the development of biologically active soil is a critical underpinning to development of productive and protective agroecosystems. Scientists can assist farmers in testing out this idea by characterizing the agroecological processes that result from the practices that farmers believe develop a biologically active soil. Such partnerships can help discover the potential of biologically active soils to benefit agroecosystems and develop rules of thumb that will broadly guide farmers in making well-founded choices in developing biodiverse agroecosystems. University extensionists can help as well, by assisting farmers in developing individual and cooperative learning skills. Such new working arrangements among farmers, scientists, and extensionists—and others awaiting discovery—are likely to be crucial to rapid development of effective biodiverse farms (see chapter 1 on the partnership among farmers, scientists, and others developed by the Land Stewardship Project in Minnesota). The Landcare movement in Australia, a rapidly evolving network of over four thousand watershed stewardship groups (Campbell 1998) is an extremely promising example of the new social arrangements we need.

Conclusions

Agroecological restoration offers a great many benefits to our commonwealth ranging from the quality and safety of food to the integrity of the landscapes that are at the heart of our physical and spiritual wellbeing. Certainly, making farms that are both protective and profitable—and resolving the contradictions between these ideals—must be central to the work of agroecological restoration. The development of more fruitful ecological partnerships through enlightened design, artful husbandry, and fleet learning offers a fundamental basis for agroecological restoration. Such a restoration will require a new synthesis of art, craft, and science formed by contributions from many different ways of knowing. This synthesis is emerging from experimentation worldwide, with many heartening results, such as the movement to develop permaculture, a method of ecological design emphasizing deployment of biodiversity to provide multiple ecological services (Mollison and Slay 1988). However, these efforts desperately need recognition and support, particularly by established institutions of agricultural science and policy. We must argue for the absolute importance of agroecological restoration in the face of relentless promotion of simplistic, "silver bullet" biotechnological solutions to agricultural problems.

References

Altieri, M. A. 1999. "The Ecological Role of Biodiversity in Agroecosystems." *Agriculture, Ecosystems, and Environment* 74:19–31.

Andow, D. A., and D. P. Davis. 1989. "Agricultural Chemicals: Food and Environment." Pp. 191–234 in *Food and Natural Resources*, edited by D. Pimentel and C. W. Hall. Academic Press, New York.

Bugg, R. L. 1992. "Using Cover Crops to Manage Arthropods on Truck Farms." *HortScience* 27:741–745.

Campbell, A. 1998. "Fomenting Synergy: Experiences with Facilitating Landcare in Australia." Pp. 232–249 in *Facilitating Sustainable Agriculture*, edited by N. G. Röling and M. A. E. Wagemakers. Cambridge University Press, Cambridge.

Carter, M. R., E. G. Gregorich, D. W. Anderson, J. W. Doran, H. H. Janzen, and F. J. Pierce. 1997. "Concepts of Soil Quality and Their Significance." Pp. 1–19 in *Soil Quality for Crop Protection and Ecosystem Health*, edited by E. G. Gregorich and M. R. Carter. Elsevier, Amsterdam.

Crossley, D. A., G. J. House, R. M. Snider, R. J. Snider, and B. R. Stinner. 1984. "The Positive Interactions in Agroecosystems." Pp. 73–82 in *Agricultural Ecosystems—Unifying Concepts*, edited by R. Lowrance, B. R. Stinner, and G. J. House. John Wiley, New York.

De Mazancourt, C., and M. Loreau. 2000. "Grazing Optimization, Nutrient Cycling, and Spatial Heterogeneity of Plant-Herbivore Interactions: Should a Palatable Plant Evolve?" *Evolution* 54:81–92.

Goodnight, C. J., and L. Stevens. 1997. "Experimental Studies of Group Selection: What Do They Tell Us about Group Selection in Nature?" *American Naturalist* 150:S59–S79.

Herre, E. A. , N. Knowlton, U. G. Mueller, and S. A. Rehner. 1999. "The Evolution of Mutualisms: Exploring the Paths between Conflict and Cooperation." *Trends in Ecology and Evolution* 14:49–53.

Holt-Gimenez, E. 1999. "The Campesino á Campesino Movement: Farmer-Led, Sustainable Agriculture in Central America And Mexico." Pp. 297–314 in *The Paradox of Plenty: Hunger in a Bountiful World*, edited by D. Boucher. Food First Books, Oakland, Calif.

Hooker, J. E., and K. E. Black. 1995. "Arbuscular Mycorrhizal Fungi as Components of Sustainable Plant-Soil Systems." *Critical Reviews in Biotechnology* 15:201–212.

Johnson, N. C. 1993. "Can Fertilization of Soil Select Less Mutualistic Mycorrhizae?" *Ecological Applications* 3:749–757.

Kinkel, L. L. 1997. "Microbial Population Dynamics on Plant Surfaces." *Annual Review of Phytopathology* 35:327–347.

Letourneau, D. K. 1998. "Conservation Biology: Lessons for Conserving Natural Enemies." Pp. 9–38 in *Conservation Biological Control*, edited by P. Barbosa. Academic Press, San Diego, Calif.

Matson, P. A., W. J. Parton, A. G. Power, and M. J. Swift. 1997. "Agricultural Intensification and Ecosystem Properties." *Science* 277:504–509.

Miller, R.M., and J. D. Jastrow. 1990. "Hierarchy of Root and Mycorrhizal Fungal Interactions with Soil Aggregation." *Soil Biology and Biochemistry* 22:579–584.

Mollison, B., and R. M. Slay. 1988. *Permaculture: A Designers' Manual*. Tagari, Tyalgum, Australia.

Perry, D. A. 1995. "Self-Organizing Systems across Scales." *Trends in Ecology and Evolution* 10:241–244.

Perry, D. A., M. A. Amaranthus, J. G. Borchers, S. L. Borchers, and R. E. Brainerd. 1989. "Bootstrapping in Ecosystems." *Bioscience* 39:230–237.

Pimentel, D., L. McLaughlin, A. Zepp, B. Lakitan, T. Kraus, P. Kleinman, F. Vancini, W. J. Roach, E. Graap, and W. S. Keeton. 1991. "Environmental and Economic Effects of Reducing Pesticide Use: A Substantial Reduction in Pesticides Might Increase Food Costs Only Slightly." *Bioscience* 41:402–409.

Somers, N. 1998. "Learning about Sustainable Agriculture: The Case of Dutch Arable Farmers." Pp. 125–133 in *Facilitating Sustainable Agriculture*, edited by N. G. Röling and M. A. E. Wagemakers. Cambridge University Press, Cambridge.

Swift, M. J. 1997. "Biological Management of Soil Fertility as a Component of Sustainable Agriculture: Perspectives and Prospects with Particular Reference to Tropical Regions." Pp. 137–159 in *Soil Ecology in Sustainable Agricultural Systems*, edited by L. Brussard and R. Ferrera-Cerrato. CRC Press, Boca Raton, Fla.

Swift, M. J., and J. M. Anderson. 1993. "Biodiversity and Ecosystem Function in Agricultural Systems." Pp.15–41 in *Biodiversity and Ecosystem Function*, edited by E. D. Schultz and H. A. Mooney. Springer-Verlag, Berlin.

Swift, M. J., J. Vandermeer, P. S. Ramakrishnan, J. M. Anderson, C. K. Ong, and B. A. Hawkins. 1996. "Biodiversity and Agroecosystem Function." Pp. 261–298 in *Functional Roles of Biodiversity: A Global Perspective*, edited by H. A. Mooney, J. H. Cushman, E. Medina, O. E. Sala, and E. D. Schultz. John Wiley, New York.

Thompson, J. N. 1998. "Rapid Evolution as an Ecological Process. *Trends in Ecology and Evolution* 13:329–332.

Tivy, J. 1990. *Agricultural Ecology*. Longman Scientific & Technical, Harlow, Essex, U.K.

Wilkinson, D. M. 1998. "The Evolutionary Ecology of Mycorrhizal Networks." *Oikos* 82:407–410.

Wilson, D. S. 1997. "Biological Communities as Functionally Organized Units." *Ecology* 78:2018–2024.

Chapter 12

Conservation and Agriculture as Neighbors

Judith D. Soule

In 1981, as I emerged from a Ph.D. program in ecology, I sought real-world applications of the principles of ecosystem functions and ecological interactions I had been studying. Two areas of particular interest emerged: conserving biodiversity and enhancing sustainability of agriculture. I spent the next twenty years bouncing back and forth between them, often wondering why it seemed so hard to bring the two together, despite a common dependence on ecological processes for long-term success—water and nutrient cycles, soil development, herbivory, predation, competition, migration, and gene flow. Most recently, I have been immersed in the biodiversity conservation world, working for The Nature Conservancy (TNC) from 1990 to 2000, a period that allowed me to witness a shift in focus within this organization.

By 1990, The Nature Conservancy, a leading player in the realm of private conservation, recognized the need to expand the scope of its conservation efforts from the site to the landscape, from species to ecosystems, and to engage other landowners in cooperative efforts to achieve its ambitious conservation goals (Sawhill 1990). This expanded scope forced The Nature Conservancy to consider the ecological and social context of its projects, and for many the context is agriculture. The dominant perception in the organization is that agriculture—used very broadly here to include both crop and livestock operations—is generally a threat to biodiversity conservation. This perception is reinforced by an analysis of the threats faced by species listed as threatened, endangered, or candidates under the federal Endangered Species Act, which showed that agriculture impacts a higher percentage of this group than any other threat—38 percent were affected by cropping and 22 percent by livestock grazing (Stein et al. 2000). Prominent ecologists also place agriculture among the major causes of habitat loss leading to loss of biological diversity (e.g., Ehrlich 1988).

The perception of threat goes both ways, with farmers usually assuming that conservationists are primarily interested in taking land out of production or putting restrictions on agriculture. As Charles Loop, a southern Texas farmer, remarked about The Nature Conservancy's new Southmost Preserve, which sits among cotton fields and citrus orchards, "There are some bad feelings between farmers and the conservation community. [Conservation organizations] have purchased lots of land and moved some agriculture out. TNC has their work cut out for them. . . ." Despite the truth in both sides' perceptions, The Nature Conservancy realized that in many cases agricultural land uses were preferable to other uses. "[Agricultural] landowners share an ethic of land stewardship that although sometimes defined differently from our own, shares the common desire to take good care of the land," stated Vosick and Cash in a 1996 TNC report.

In this context, I sought out examples of projects where The Nature Conservancy is working on conservation in an agricultural setting to see to what extent this organization has been able to establish common ground and positive outcomes with its farm and ranch neighbors. The eight projects I investigated fell with surprising neatness into four categories of fundamental relationships and three sets of strategies for working with the agricultural community. They illustrate many of the nuances and complex challenges of blending conservation of biodiversity with agriculture.

Fundamental Relationships

The relationships between agriculture and conservation of biodiversity run the gamut from one-sided effects to mutual harm or mutual benefit. For example, one-sided relationships might involve agricultural threats to conservation via exports of nutrients and toxins that alter nutrient cycles or food-chain relationships in adjacent conservation areas, or they could entail threats to agriculture from conservation sites via agricultural pests harbored in native ecosystems. Conversely, conservation areas may harbor pest enemies or needed pollinators that benefit neighboring agriculture, and agricultural systems can at times provide habitat, food, or disturbance regimes that support conservation of native species or vegetation. Many times the fundamental conflicts between biodiversity conservation and agriculture lie in these sorts of ecological relationships, though socioeconomic factors frequently come into play also. The following examples illustrate a variety of fundamental relationships between agriculture and conservation, both ecological and socioeconomic. None seem to illustrate the mutual harm case, although any of them could potentially slip into that category if things went wrong.

Agriculture's "Exports" Threaten Conservation Targets

Three TNC projects in the Great Lakes and Ohio River drainages share similar conservation targets: French Creek in western New York and Pennsylvania, Upper St. Joseph Watershed in southern Michigan and northern Indiana, and Big Darby Watershed in south-central Ohio. Water quality and flow regimes in each of these river systems are important for supporting exceptionally diverse communities of rare fish and mussels, groups that are high on the list of conservation priorities according to two recent analyses (Stein et al. 2000, WWF 2000). All three projects are set in agricultural landscapes, but the differences in the type of agriculture, productivity of the land, and regional socioeconomics create unique outcomes in each case.

Susan McAlpine, project director for The Nature Conservancy's French Creek Project, can rattle off statistics about Holstein cattle without pause. "They average 1,400 pounds in weight and in one day produce 80 pounds of waste—manure and urine—one hundred pounds of milk, and drink 40 gallons of water. They're big machines!" she says. Dairy farming is the predominant land use in the watershed and the predominant source of nutrient and sediment loading in the stream. French Creek, with the richest fish and mussel community in the northeast, originates in the hills of southwestern New York State and runs south through western Pennsylvania to the Allegheny River. The 80 square miles of the creek's headwaters that lie in New York state are in a relatively isolated rural area where cows outnumber human residents two to one and 30 percent of its barnyards lie within 150 feet of a stream. When McAlpine arrived in the watershed in 1993, most of the farmyards and pastures had no fences along the stream and few had adequate barnyard nutrient management. Seepage and storm flushing of barnyard materials and soil into adjacent streams degraded water quality and in-stream habitats, presumably posing serious threats to the aquatic communities that The Nature Conservancy wished to protect.

The Upper St. Joseph Watershed project in northern Indiana and southern Michigan started about the same time as French Creek, in 1992, when Larry Clemens was hired by The Nature Conservancy's Indiana chapter as project manager. Two of the upper tributaries to the river, Fish Creek in Indiana and the East Branch of the West Fork in Michigan, are home to the most diverse suite of rare freshwater clams and fish in northern Indiana and Ohio, and perhaps in the Great Lakes watershed. Here, corn and soybeans are the dominant crops, and the challenge is to limit movement of sediments, nutrients, and pesticides from crop fields into the streams.

In south-central Ohio, the Big Darby Watershed sits in a fairly flat landscape that was originally pocketed with wetlands, 90 percent of which have been converted to farmland. The Darby's mussel community, with thirty-eight species, is far richer than that of either French Creek or the St. Joseph. The cropland is richer also. The Darby farms, like those in the Upper St. Joseph, grow corn and soybeans, but the Darby farmers have generally been able to make more money farming than putting their land into the Conservation Reserve Program (CRP) at the $80-per-acre government lease rate. The St. Joseph headwater counties, by contrast, have the highest concentration of CRP lands in the eastern corn belt. Again, farm run-off is the primary concern for conservation of the rich aquatic community.

In all three projects, the fundamental relationship between conservation and agriculture is shaped by the awareness that agricultural practices cause ecological threats to the conservation targets. Conservation has everything to gain by changing agriculture, but agriculture has little, if anything, to gain directly from the preservation of the conservation targets. The one-sided benefits in this sort of relationship make it especially tough to establish common ground between farmers and conservationists. Duly noting this skewed relationship, The Nature Conservancy sought ways to provide some benefit for the farmers while still achieving its conservation goals. In two of the three projects, this translated to financial incentives for farmers to change practices.

In the French Creek and Upper St. Joseph watersheds, the focus of TNC actions is to promote adoption of improved nutrient management (e.g., stream-bank protection, nutrient management plans, and barnyard management practices in the former case, and conservation tillage in the latter) by as many farmers as possible. The Nature Conservancy addressed the economic barrier to change head-on by helping to defray farmers' transition expenses and providing evidence that the new practices would save the farmers money. Several factors were important in The Nature Conservancy's ability to build credibility and trust in the project areas: successful partnerships with long-trusted groups such as the Natural Resources Conservation Service (NRCS), county Soil and Water Conservation Districts, and Cooperative Extension programs; a long-term presence in the area; actual faces for folks to relate to; outside money spent locally; and no major blunders made. At French Creek, despite a farming community characterized as skeptical of conservationists and slow to adopt new practices, best management practices (BMPs) have been adopted on 50 percent of the streamside farms and 30 percent in the watershed as a whole. On the Upper St. Joseph, acreage of corn and soybeans grown with conservation tillage rose from about 33 percent in 1992 to 75 percent of the soybeans and 50 percent of the corn in 2000.

The Darby project approached things differently. Rather than offering financial incentives, it sought to address sociocultural barriers to change by facilitating farmer-to-farmer learning and community organization around the conservation values of the Big Darby river. The "Darby Partnership," a group of about fifty agencies, organizations, citizens, and farmers, promoted communication about the value of the river and identified how the various partners could contribute to its conservation. "Operation Future Association," with leadership from Ohio State University Extension, provided farmers opportunities to learn from one another how to successfully use conservation tillage. The Darby's reliance on partnerships apparently paid off. Conservation tillage in the watershed has increased 147 percent, over 400 acres of trees have been planted along the creeks, and nutrient and pest management plans have been applied to over 14,500 acres. NRCS data from tillage-transect studies showed a 35,000-ton reduction in sediments reaching the Darby creeks from agricultural operations.

Although the three projects have figures summarizing changes in agricultural practices, it is not entirely clear how much change can be attributed to the projects and how much is unrelated regional change. Nor is it easy to attribute conservation benefits to these changes, since biological baseline and change data are also scant (though studies are now underway). It is probably too early to detect changes in the mussel communities, at any rate, because they are long-lived as adults, with twenty- to thirty-year life spans. Just as gaining trust within the local community and instituting changes in farm practices are slow processes, the ecological healing processes are also slow. These are projects built on the conviction that over the long term they will make a difference.

In the absence of definitive data, the instincts of all three project managers are that the progress so far is not enough to assure their conservation goals. The Darby is still losing mussel species, and project manager Laura Belleville remains concerned that the changes in farming practices are not necessarily permanent, and is doubtful that the farming community has made a long-term commitment to conservation. Farmers adamantly opposed a recent proposal for a federal wildlife refuge on the Darby despite a "willing seller only" clause and a provision for over 20,000 acres in farmland preservation. Belleville explains that several socioeconomic factors contributed to farmer opposition to the refuge: the uncertain future of global soybean markets; the fact that few of the younger generation are opting to farm; the need to retain options to finance their retirement. Combined, these factors outweighed farmers' desire to see farmland stay in the farming.

Clemens is concerned that some farmers in the St. Joseph watershed are riding the new wave of conservation tillage and then converting back to conventional farming when they meet with frustration. "They're just trying to make ends meet," he sympathizes, yet he realizes that pointing farmers to sustainable methods doesn't provide permanent protection. On the other hand, Clemens has also seen a general shift in attitude among farmers during his tenure with the Upper St. Joseph project. Their early consensus was that permanent protection was undesirable, but gradually farmers have begun to contact The Nature Conservancy about obtaining protection for their land. Clemens attributes the shift to a growing recognition that the best use of the marginal farmland in the floodplain is in forests that provide buffering against floods. He observes, "The older the landowners get, the more they want to do conservation-minded things. It is almost like repentance. . . . They've spent fifty years clearing and farming the land, and now they want to give something back to the land. Most of the current residents along the river corridor grew up there and want their kids to be able to enjoy it like they did." Over 1,000 acres have been planted to trees to reestablish forested land. Says Clemens, "If we could go back one hundred years, we should have left this land in forest. It would have been worth more in forest products than in crops."

At French Creek, it is not so much the lack of permanence of adopting best management practices that troubles McAlpine, but rather the slow progress and the likelihood that BMPs alone are inadequate to protect the creek's biodiversity. A recent study about barriers to streamside revegetation revealed the cultural complexity of change. The study found that farmers liked the look of a cleared stream corridor, and some considered the brush a hazard to calves. They had invested hard labor in clearing out the vegetation and brush along the streams and were not eager to let it grow back to a "messy" state. It will take a much higher number than 30 percent or even 50 percent of farms instituting BMPs to see significant water-quality concerns abate. "Improvements along one farm's streamside can't fully compensate for upstream impacts," says Reuben Goforth, aquatic ecologist with the Michigan Natural Features Inventory, Michigan State University Extension, who conducted research on the ecology of French Creek in relation to landscape patterns. New regulations of waste management for large herds may help speed up adoption of BMPs, but, ultimately, it will also be necessary to restore the hydrology of the creeks to less-variable flow regimes, and that will require wetland restoration and an increase in permanent vegetative cover in the watershed.

The sometimes ephemeral nature of changes in farm practices raises a fundamental dilemma for The Nature Conservancy, which is looking for permanence, for insurance to its contributors that their dollars are funding actions that are forever and not just for the fifteen-year time frame of many farm programs. David Weekes, then Midwest Division Vice President for The Nature Conservancy, worries that in all three of these project areas, despite apparent conservation gains from changed agricultural practices, the land largely has no permanent protection status: "It makes no difference if a farmer used conservation tillage; once he sells out to a developer, most likely, conservation loses." And thus, in all three projects, after nearly a decade on the ground, strategies are shifting or expanding from a primary focus on "changing agriculture" one farm at a time, to a more complex set of strategies: shifting attention to the urbanizing portion of the Darby watershed; supporting new holistic sustainable agriculture curriculum development; promoting intensive rotational grazing; facilitating land and easement purchase; and looking for permanent, wide-reaching solutions for conservation and agriculture.

Agriculture and Conservation Compete for Land and Water

The Platte River/Rainwater Basin project in Nebraska and the Clive Runnells Family Mad Island Marsh Preserve on the Texas Gulf Coast have a few things in common (although certainly not geography). Both projects have protection of migratory birds and restoration of natural habitat as primary goals. Both are affected by issues of water allocation. And they both illustrate a fundamental relationship between agriculture and conservation goals different than that seen in the projects described in the previous section.

Jim Bergan, Texas TNC's Director of Science and Stewardship and original project director for the Mad Island Marsh Preserve, is describing some of the benefits for farmers of using a second crop flooding in their rice production. I'm struggling, as though looking through a stereoscope with different pictures under each lens, trying to bring it into three dimensions but finding no points of congruence between the two images. Picture a group of your typical Audubon or Natural Areas Association field trip participants—mostly over fifty, or under twenty-five—fit, leathery folk with hiking boots, field guides poking out of various pockets and packs, binoculars slung around their necks. Now picture a Texas rice farmer standing by his Texas-sized enclosed-cab combine, collecting fees, as participants climb aboard for an early morning tour around the farm's rice fields. The birders are excited about the possibility of seeing perhaps a dozen of the notoriously cryptic yellow rails flush. The farmer may also enjoy the sight of the birds rising from his fields but has the added pleasure of some income that is relatively independent of weather, pests, and bank loans.

In 1991, when Bergan took the job at Mad Island, the preserve, though set in the midst of productive rice farmland (or former wetlands, depending on your viewpoint), contained only natural habitats—a mix of near-coast freshwater wetlands, estuaries, bottomland forest along Texas' Colorado River, wet prairie, and other types of wetlands. These natural habitats, especially the freshwater wetlands, which are among the most degraded ecosystems along the Gulf Coast, and the waterfowl, shorebirds, and marsh birds that depend on them, are the primary conservation targets of the preserve. Competition for freshwater from cities and industry threatens to starve the wetlands and estuaries of freshwater, but rice farming, at least, returns freshwater to the wetlands. To The Nature Conservancy, rice farms were the preferred land use for their neighbors. In 1993, they added 1,500 acres of rice fields to the core preserve. Only about 175 acres of low-productivity farmland have been removed from farming for restoration to permanent wetland and wet prairie habitats. The rest is leased to the previous farmers to set up a demonstration area for bird-compatible growing practices. The farmers helped The Nature Conservancy to develop rotations that keep about one-quarter of the land in production at any time and that include flooding the second rice crop to simulate wetland habitat for the birds. (See chapter 13 for an extensive description of a similar wetland-cropland rotation in northern California.)

About the same time that Clemens was hired in Indiana, TNC Nebraska hired Brent Lathrop to start the Platte River/Rainwater Basin project. The project's name refers to two areas, both vitally important for migratory birds: the Platte River in central Nebraska and the Rainwater Basin. The latter refers to a seventeen-county 4,200-square-mile area that has similarities to the "prairie potholes" of the Dakotas and the Canadian Great Plains, the difference being that the Nebraska version was carved by wind, rather than by glaciers. Only about 12,000 to 15,000 acres out of the original 100,000 acres of the basin wetlands remain. Together, the Rainwater Basin and an 80-mile stretch of the Platte River near Grand Island, Nebraska, support about 10 million ducks and geese and 500,000 migrating sandhill and whooping cranes each year. An abundance of migrating grassland birds use both areas as well. The Nature Conservancy owns about 3,000 acres along the Platte. Irrigated corn dominates the land along the river today, but it is a relatively new entrant into the ecology of the Platte River system. Through the 1960s, most of this land remained in native prairie and was used for grazing cattle. In the 1970s, the cattle ranching gave way as the prairie was turned under to plant corn for export commodity programs. The acreage of prairie plowed under during that decade rivaled the total lost during settlement.

The Nature Conservancy intends to restore as much of the native prairie along the river as possible. On TNC-owned lands, however, restoring prairie doesn't mean eliminating agriculture; it means substituting cattle-grazing land and hay fields for irrigated corn. Working with the farmer leasing the TNC land, corn is phased out over about three years as prairie is restored. The farmer can continue to use the land for grazing or hay production and on the side make some income from a native seed collection business that supplies organizations doing prairie restorations, or from providing blinds and access for the many bird enthusiasts who visit each spring to see the awesome aggregation of cranes and geese.

In both of these projects, the primary relationship between agriculture and conservation is competition for space on the landscape. But in both cases an element of mutual benefit also exists. Agriculture can help restore or replace lost habitat (grazed native prairie, or rice field "wetlands") and conservation can provide nature-based enterprises for added farm income, though the economic and conservation outcomes are not yet well known.

For the farmer, the cost-to-benefit ratio of the changes in agricultural systems is unclear, though studies are underway at both sites. A University of Nebraska study is investigating the profitability of grazing cattle compared to raising corn, with and without government subsidies and with different mixes of grasses. In Texas, a Kellogg Foundation grant to The Nature Conservancy will be used to measure costs and benefits to farmers of the bird-friendly practices, while Bergan seeks funding to investigate how the late flooding system affects pests and pesticide use.

As for outcomes for conservation, neither project has much data to support conclusions about successes to their credit other than the acreage restored to native habitat. Christmas bird counts conducted at Mad Island have been especially high for the past three years, but they do not have data from the project's inception, and the counts cover the entire diverse landscape of the preserve, not just the simulated rice field wetlands.

Unlike the French Creek and St. Joseph Watershed projects, these projects have depended primarily on demonstrating new practices on TNC-owned land rather than working to convince surrounding farmers to change their practices. Despite this lack of attention, or perhaps because of it, both project leaders have noticed some neighbors experimenting with similar systems on their own land. Both also recognize the need for change to extend beyond TNC ownership to accomplish their goals and are searching for new ways to make change more attractive.

Bergan believes he's found the carrot for increasing the use of wetland-mimicking farm practices. "It all comes down to water. Agriculture is competing with corporations and municipalities for water, and agriculture can't compete," says Bergan. He is hoping to add weight to the agricultural claim on water by quantifying the wildlife values of the water used in the integrated rice-wetlands system. Water is used multiple times in this system—for crop production, to charge wetlands, and ultimately as freshwater input into the tidal wetlands of the gulf, where it is essential for supporting oysters and the entire diverse nearshore marine community. With strong indications that the

Environmental Protection Agency (EPA) will provide the stick and begin regulating rice field tailwater within the next few years, Bergan has sought research funding from the rice commodity associations without success. He has managed to conduct a pilot study on effects of rice field tailwater in coastal ecosystems, but it needs replication and expansion at an estimated cost of about $300,000 to provide scientifically valid results. So far no impacts from farm pesticides have emerged, but it is not so clear whether or not there is a nutrient effect.

Although water rights are also an issue in the ecology of the Platte River, Lathrop's focus is on putting in place support to help make the shift from corn to cattle easier. Even if the 1970s don't seem that long ago to this author, it is apparently long enough that much of the Platte River farm community's knowledge about grass and cattle management has been lost. So The Nature Conservancy is hiring a grass and cattle specialist to focus on sustainable grazing on restored grasslands and is considering developing grazing cooperatives, or "grass banks." Grass banks would strike an agreement with farmers to manage their cattle in ways that sustain the natural values of the prairie. In return, if a farmer needed more forage in an especially dry year, for example, they could bring cattle onto TNC land to graze temporarily. Like the leaders of the three river projects, Lathrop is grasping for ways to make enduring impacts and he sees partnership with the Prairie Plains Resource Institute to develop environmental education opportunities as the key. The ecological problems affecting the Platte River are much bigger than what he can address through restoring prairie acre by acre. They involve overallocation of the river's water, and controlled flows that change the processes which create and maintain the river channel. The former braided river is nearly gone and so too is crucial habitat for the migrating birds. As Lathrop says, "It's the next generation where [our efforts] will start showing," and this is where he finds hope.

In contrast to the three river projects, the conservation targets for Mad Island and Platte River have clearer benefits to agriculture. Prairie can be used to raise beef and provide side income from seed harvest and ecotourism. Rice fields can be an analogue for wetlands and perhaps secure the farmer's claim on water while providing ecotourism income on the side. Mussels and small fish in a creek don't have a parallel benefit to the dairy farmer. Important "ifs" remain: the benefits will be realized only if those with power over water rights accept that conservation values provide valid claims on water; if the recreation business can provide enough income without ecological damage to make a difference; if the farmers can make a living using the practices beneficial to birds; if there is a market for grass-fed beef; if the tailwater from the rice lands is clean enough to not upset the ecology of the wetlands. Investing in answering as many of these questions as possible would be worthwhile.

Agriculture Assists with Stewardship of Conservation Areas

Another variation on the relationship between conservation of biological diversity and agriculture occurs when agriculture can offer an analogue of a natural ecosystem process that is missing or suppressed in the native system. The prairie fens of southern Michigan and the Red Canyon Ranch in Wyoming illustrate how this relationship can work. In both cases, the primary focus of conservation is restoration and stewardship of natural communities/ecosystems.

In southern Michigan, prairie fens are a rare natural community with peat soils, open grasses and sedges, and scattered shrubs. They are habitat for a number of rare species, including the Mitchell's satyr butterfly, listed as endangered under the federal Endangered Species Act. Fens form where calcium-rich groundwater seeps from the base of steep slopes and usually occur in complexes of oak savanna, prairie, wet meadows, and shrub swamps (MNFI 1997). Across southern Michigan, prairie fens are rapidly succumbing to invasion by shrubs because of a combination of altered hydrology and suppression of natural disturbances. Land use around most fens is largely agricultural (soybeans or livestock) or residential. Where tiles and drainage ditches from surrounding fields empty or continue into a fen, the sheet flow from groundwater seepage that makes a fen a fen is disrupted. Where crop fields or residential septic fields abut a fen, aggressive cattails and exotic phragmites and purple loosestrife often take hold, signaling nitrogen leaking from the fields. At The Nature Conservancy's Ives Road Fen Preserve, staff and volunteers have spent countless hours beating back the shrubs, filling ditches to repair hydrology, and conducting prescribed burns. The Nature Conservancy's Grand River Fen Nature Preserve, by contrast, has little shrub encroachment and requires much less stewardship. The difference may relate to historic farming practices, which until recently included stubble burns that ran down into the fen at Grand River Fen only, and tiling at Ives Road Fen only. At present, compared to restoring hydrology and fire management, reduction of nitrogen inputs from farm runoff is a secondary concern, and The Nature Conservancy is not addressing agriculture directly.

However, an unexpected incident at another fen occupied by Mitchell's satyrs, suggests that the relationship between fens and farming may deserve more attention. When biologists Daria Hyde and Matt Smar from the Michigan Natural Features Inventory visited the "Kimberly" Road fen (name changed for confidentiality) in the spring of 2000 to search for Mitchell's satyr larvae, they were alarmed to discover the fen grazed nearly bare by the farmer's 100-head dairy herd. Only on the hummocks did some tufts of sedges, the larval food source, remain, and Hyde and Smar found no larvae. The farmer indicated that she let the cows into the fen each year, where they would rest under the oaks on the higher slopes, but that a dry spring may have allowed the cows access to usually soggy parts of the fen. When biologists returned to the site in July to census adult butterflies, to their delight, they found nearly one hundred adult satyrs, many more than in previous counts at the site. Hyde pointed out that the meaning of the high count is not clear. Were the butterflies easier to find because of the scant cover? Was it coincidentally an exceptionally good satyr year? Or, did the grazing stimulate new, more nutritious growth of sedges, dramatically increasing the satyr's larval survival rate? Surprisingly, the one small study to date on Mitchell's satyr larval food preferences found that larvae fed more often on mature leaves than young ones (Jennifer Syzmanski, personal communication), so the answer is not obvious. Next year's census may help answer these questions, but in the meantime, it seems worth investigating the grazing history of fens.

Reflecting on this story, TNC Michigan's land steward, Jack McGowan-Stinsky cited cattle bones found at Ives Road Fen as evidence of past grazing. TNC stewardship ecologist Christopher Clampitt and Hyde each recalled seeing old fences extending into several fens and at both Ives Road Fen and Grand River Fen, abandoned rolls of fencing wire have been found—the old flat-barbed style at Ives Road Fen. Apparently at some point in the past, farmers allowed their livestock access to the fens, a practice that made some sense to Clampitt: "Fens are rich in calcium and magnesium, though low in nitrogen. Their sedges and grasses would make nutritious forage, especially for dairy cattle." Local lore indicates that deer inhabiting these fens grow extra-large antlers. "This is where the big racks come from," noted McGowan-Stinsky, an observation consistent with studies showing that minerals in the diet are a factor in antler growth (Bill Moritz, Michigan Department of Natural Resources wildlife biologist, personal communication). Clampitt and McGowan-Stinsky were willing to consider that grazing might play a beneficial role in fen management, although they favored fire management, citing soil disturbance and added nitrogen associated with grazing as providing a foothold for purple loosestrife invasion. Scientists differ in their assessments of the relative benefits and hazards for insects of fire versus grazing, and at least one case where both are necessary for a butterfly population has been described (Ausden and Treweek 1995, Schultz and Crone 1998, Pyle 2000). Prior to European settlement, the region was subject to fire management by Native Americans and periodic grazing by native elk and bison. Grazing could prove to be a useful addition to fen stewardship tools by keeping shrub invasion low, perhaps promoting Mitchell's satyr larval growth, and simultaneously providing nutritious forage for livestock. This would be, fundamentally, a relationship of mutual benefit between conservation and agriculture.

If the southern Michigan fens *suggest* a mutually beneficial relationship with agriculture, The Nature Conservancy's Red Canyon Ranch in Wyoming *shouts* it. For Bob Budd, who runs the Red Canyon Ranch, the conservation targets and the agricultural operation, in this case cattle ranching, are virtually indistinguishable. The diverse natural communities that make up the landscape of the ranch evolved with a suite of grazers and were adapted to a disturbance regime that included grazing and fire. Today, the ranch simulates the roaming habits of herds of elk, bison, wild sheep, deer, and pronghorns by moving the cattle frequently around the ranch, grazing different sections in different seasons in subsequent years. This system costs more for management than a "free-range" system. It takes a rider whose full-time job is to move the cattle around the ranch. But the system's payback comes from reduced losses, less-extensive fencing, and allowing the ranch to run a larger herd—50 percent larger—while the ecosystems look healthier each year. Monitoring at hundreds of photo points over the five-year life of the project has documented healing of the riparian corridors. Bird counts taken in conjunction with the Wyoming Game and Fish Department have documented, in addition to raptors, an increasing diversity of Neotropical migrants across the 35,000-acre ranch (a mix of TNC-owned and federal land). He advocates using birds and fish as the primary monitoring tools; the birds seem to tell you about the diversity of habitats and the fish tell about the health of the riparian zone.

The focus of the Red Canyon Ranch project is on the TNC-owned ranch itself, not on the management of their neighbors' operations. "We don't hope for anything from neighbors except for them to stay in business," says Budd. Ranches are better for the land than the alternative, subdivisions and other forms of habitat fragmentation, which even in Wyoming are beginning to sprout up. He describes relationships with his ranching neighbors as being based on a two-way exchange and learning process. Many of the ideas tried on the ranch were suggested by neighbors, or tried elsewhere first, so Budd seeks no credit if others in the area also use some of the same innovations he's using. He readily acknowledges that he has the luxury of not having to attend to the bottom line to the same degree his neighbors do, a luxury that makes it far less risky for him to experiment.

The potential for mutually beneficial relationships in the fens and the ranch derives from the assumption that cattle can provide disturbance analogous to that provided by native herbivores now absent from the ecosystems. The Platte River/Rainwater Basin prairie restoration project also uses this relationship to maintain restored prairie. Opinions differ among scientists as to the conservation value of grazing and whether cattle are an acceptable ecological analogue to native herbivores. In the United States, conservationists tend to favor fire management over cattle grazing, blaming the latter for reduction in diversity and spread of invasive exotic plants (debate and references summarized in WallisDeVries 1998). Other recent studies have shown that grazing can have positive or neutral effects on ecosystems and diversity and that large-scale influences such as serious drought or variations in soils have a much greater effect on species composition than grazing does (McNaughton 1993, Biondini et al. 1998, Stohlgren et al. 1999, Schuman et al. 1999). In Europe, there is wide acceptance of the use of grazing to attain conservation goals in native grasslands with a less favorable view of fire management (WallisDeVries 1998, Ausden and Treweek 1995). Studies have shown that fire and grazing affect different plants differently, suggesting that some combination of both may lead to the greatest plant diversity in grasslands (e.g., Hatch et al. 1999). Budd is convinced that both are critical in the ecosystems on his ranch. He related a conversation he'd had with some "old-timers" about changes to the country that have occurred over the past fifty years. They all agreed—there was more water fifty years ago. They talked about springs that had dried up and now-intermittent streams that used to run year round. Budd theorizes that this relates to fire suppression, reasoning that the resulting increase in conifer and brush cover may draw enough water from the soil to dry up springs and seeps, perhaps explaining many of the woes that are often blamed on overgrazing. He points out that the ecosystems of the West evolved with a variety of disturbances, including fire, grazing, drought, flood, disease, and herbivore population fluctuations, a combination that created a highly varied landscape, including a range of successional stages. Many of the species that are now considered vulnerable, such as sage grouse, required a variety of habitats to complete their life cycle. This may also hold true in the fens of Michigan, where Hyde noted that the male satyrs were seen in the open parts of the fen while the females stayed close to the shrubs. It suggests that for disturbance-adapted ecosystems, using a varied set of disturbance agents, such as fire *and* grazing, may be of great benefit when applied to natural areas. It may be a general truth that a set of varied stewardship tools (e.g., both fire and grazing) which mimic historic disturbances is likely to provide the greatest benefit for biodiversity conservation in natural areas (Landres et al. 1999).

Conservation Areas Harbor Threats or Benefits for Agriculture

It turns out that the possibility of using grazing as a stewardship tool is only part of the story of farms and fens. In the summer of 2000, Asian soybean aphids (*Aphis glycines*) were detected for the first time in the United States across the north central Great Lakes states, including the region where fens occur. According to Doug Landis, Michigan State University entomology faculty member, in its native China, this aphid inflicts yield losses of up to 28 percent, so its introduction into North America is viewed with concern. In Wisconsin, yield losses of 10 to 12 percent were recorded in 2000. This aphid's winter host in China is a species of buckthorn (*Rhamnus davurica*) that is uncommon in the Great Lakes region. However, two European species, common buckthorn (*Rhamnus cathartica*) and glossy buckthorn (*R. frangula*), were introduced from Europe for hedges and shelterbelts, then spread aggressively into native habitats and are now common in the region. Glossy buckthorn is the most problematic shrub invading Michigan's fens. Common buckthorn has already been shown to be an acceptable winter host for the aphid (though not yet documented in the field), but glossy buckthorn has not yet been tested and may prove to be acceptable also (DiFonzo and Hines 2001).

Here is a case where poor stewardship of natural areas could pose a threat to agriculture by fostering a serious farm pest. Addressing the stewardship issue by keeping buckthorn in check could benefit farmers by removing the soybean aphid's wintering habitat and could benefit the rare species of fens by keeping an appropriate mix of open and shrubby habitat. Wouldn't it be interesting if that stewardship included a little grazing?

The role of natural habitat in agricultural pest control in general is a subject that is gaining more interest in the science of biological control. Landis and colleagues (2000) reviewed the literature on "conservation biological control," the termed used for providing habitat for natural enemies of crop pests in conjunction with crops to enhance effectiveness of biological control. Natural or semi-natural habitat can provide alternative food sources, shelter, alternative prey, or hosts for natural enemies of crop pests. This helps keep beneficial insect populations high enough to significantly impact pests. From the story of fens it is clear that natural habitat can also harbor the pests themselves, making proper management essential.

The potential for benefits such as those Landis described may apply to the Mad Island Marsh Preserve rice fields, and also at The Nature Conservancy's Southmost Preserve, a new preserve at the southernmost tip of Texas, and would be worth investigating in both cases. Southmost Preserve is situated in an extremely diverse corner of the United States, a tropical corner, really, at latitude 25.5 degrees. This preserve has a rich diversity of rare species of plants and animals, 7 miles of riparian forest along the Rio Grande, a *resaca*, or oxbow lake, Tamaulipan thornscrub—habitat for ocelot and jaguarundi, and one of only two stands of Mexican sabal palm in the country. It is unusual because it also incorporates active cotton and sorghum fields and a citrus orchard. Charles Loop, who has farmed this land in the past, continues to work the farm, developing practices that minimize impacts of pesticides on birds and other components of native ecosystems. The biological diversity includes an abundance of insects, including many crop pests, and pesticide use in traditional practice in the region is intense. Loop has been ahead of the curve, using integrated pest management techniques such as pheromone traps for years. It seems a perfect site for trying out conservation biological control.

The orchard may already harbor some beneficial pest control. It is the only place on the preserve that the elusive ocelot is seen, probably attracted by the cover provided by the orchard and the abundant supply of rodents that feast on fruit falls. The preserve also has natural brushy habitat for ocelots, and on the U.S. side of the Rio Grande a corridor of natural habitat runs between the river and the croplands, providing connections to other habitat. This suggests to me that as we think about how to address the tough conservation issue of providing sufficient habitat to maintain large mobile predators, such as the ocelot, that with appropriate management, some agricultural lands could figure into the picture, providing alternate habitat, food sources, shelter, or travel corridors. Landis and colleagues envision an ideal scenario for agriculture involving a "variegated landscape," with a "matrix of native vegetation in which agriculture is nested." They conclude "the encouragement of natural enemies by strategic increases in habitat diversity offers potential to align the goals of agriculture with those of nature conservation" (Landis et al. 2000).

Conservation and Agriculture as Neighbors

Most of the projects described above have been underway for eight to ten years, long enough to provide ideas about how best to create a positive relationship between conservation and agriculture as neighbors. Sifting through the stories, I find that the conservation strategies employed by The Nature Conservancy in these eight projects seem to fall into only three categories. The strategies relate to the basic ecological relationships, and a variety of socioeconomic, cultural, and policy factors influence their success. I'll call them *catalyst*, *demonstration*, and *minding your own business*.

The three river projects all employ a catalyst strategy. Given that conservation has everything to gain by changing agriculture, but agriculture has little, if anything, to gain from saving mussels and small fish, The Nature Conservancy sought ways to provide some energy, insurance against uncertainty, and interest in change among the farm community. Although successful in bringing about some changes in agricultural practices, the degree of change of heart or development of deep commitment to conservation goals varied, apparently with demographic, socioeconomic, policy, and cultural factors. For example, where the farmland was the poorest (Fish Creek), the greatest sympathy for conservation has developed. The degree of impact on the conservation targets is unclear. Good monitoring data is needed, but the project managers suspect that greater incentives to support large-scale changes in land cover in their watersheds will be needed to really achieve their goals. Yet even then, how much can be accomplished watershed by watershed? To achieve "transformative change, at a scale that matters" (in the lingo of new TNC president, Steve McCormick), farm policy, global economics, and environmental policy must be addressed.

The Mad Island Marsh and Platte River/Rainwater Basin projects use a mixture of catalyst and demonstration strategies. This seems to match the mixture of fundamental relationships also—conventional agriculture and conservation compete for land, but the "right kind" of agriculture is compatible with conservation goals so that both can be achieved simultaneously. Catalysts (e.g., ecotourism and native-seed collection) are needed to help justify change in agricultural practices, but the mutual benefits may speak for themselves through demonstration sites. Both projects have raised awareness of conservation and demonstrations have caught the interest of neighboring farmers, but both would probably benefit from more solid economic figures. As with the three river projects, the incremental gains at Mad Island and Platte River are dwarfed by the overriding issues of water allocation and agricultural and environmental policies.

Red Canyon Ranch seems to be employing the "un-strategy" of "mind your own business." The self-contained landscape of the ranch is less affected by ranching practices of its neighbors, and the agricultural use of the land benefits rather than threatens conservation goals, so the urgency to spread the news is less than for the other projects. Red Canyon relies on demonstration, mutual learning, conversation, and low-key participation in local groups and committees. All Red Canyon Ranch needs is neighbors who ranch, which points again to overriding issues of land use, water allocation, and agricultural and environmental policies that influence a rancher's ability to stay in business.

Michigan fen projects also employ a "mind-your-own-business" strategy relative to agriculture, but stewardship issues relating to grazing and crop pests may necessitate use of demonstrations and also a fourth strategy, mutual partnerships, based on abatement of mutual threats. The new Southmost Preserve has begun by using a demonstration strategy, but the potential for one-way benefits and threats suggests a mixed strategy including incentives will be required for success.

Regardless of fundamental relationships and strategies used, communication is key to success in all of these projects. Two of the project leaders commented on communication barriers between the agricultural and conservation communities. Interestingly, one seems to have learned from mistakes and the other from success. In the Darby project, despite the emphasis on talking and communication, it became clear when the federal wildlife refuge proposal emerged that communication had not been effective. The Nature Conservancy was taken by surprise when the farmers rejected farmland preservation. Belleville related some thoughts from Bob Barger, a conservation partner of The Nature Conservancy's in the Darby area who has straddled the line between academia and farming in his personal life. He sees the different learning styles employed by farmers and conservationists, in general, as barriers to communication. While conservationists tend to start from the abstract, to think of long-term consequences, and to use logic and scientific facts to learn and solve problems, farmers tend to start from their own concrete observations and experience, to think in a year-to-year time frame, and to rely on storytelling and anecdotes to learn and solve problems. Neither group has learned to speak or fully listen to the other's language.

Bob Budd's observations from Red Canyon Ranch reinforce Barger's. Budd has noticed that "ranchers talk in questions." Conversations start with "So, how much snow did you get?" "Did you read that thing in the paper about grazing . . . ?" Ranchers are careful to solicit your opinion before saying much about what they think. It is a nonconfrontational style, adaptive in a community where you may need help from your neighbor tomorrow. In contrast, conservationists tend to take the direct approach. They are more likely to tell you what you should do rather than ask you what you think, likely to look at what is wrong and ignore what is going right, be critical instead of appreciative, and impatient to get to a solution. The rancher may take longer to get around to the problem areas, but the conservationist's impatience may deny the discovery that the rancher recognizes the problems too and has some good ideas for solutions. A demonstration that a rancher can explore, contribute to, discover for him or herself is more likely to bring about change in thinking than a string of facts and figures and a logical argument why the new method is better than the old.

A couple of aphorisms may be adequate to sum up the lessons in these stories. Borrowing from *Poor Richard's Almanac:*

> *A good example is the best sermon.* The less conservation advocates talk and sell their ideas with logic and facts, and the more they demonstrate and explore them with their farm neighbors, making it a truly two-way exchange, the more likely that the neighbors will adopt conservation goals and new practices.
>
> *Don't put all your eggs in one basket*, particularly if the ribs are made of federal policies and short-term economics. Changes in agricultural practice are not reliably permanent, influenced by the economics of change, social demographics, cultural traditions and per-

ceptions, and external national or global economic policies. Conservation practitioners need to use multiple tools and find ways to make their commitment to conservation contagious within the communities that influence their conservation targets.

The stories related above reveal many opportunities for agriculture and biodiversity conservation to coexist, provide mutual benefits, and even to merge on a single piece of land. Success in finding this common ground depends on mutual respect, appropriate communication, and mindfulness of the larger issues and long-term goals. This will require that the conservation community learn to really listen to farmers and ranchers and to communicate in ways that are effective with them. It will also require attention to national farm policy and support for policies that catalyze good stewardship practices.

References

Ausden, M., and J. Treweek. 1995. "Grasslands." Pp. 197–229 in *Managing Habitats for Conservation*, edited by W. J. Sutherland and D. A. Hill. Cambridge University Press, Cambridge.

Biondini, M. E., B. D. Patton, and P. E. Nyren. 1998. "Grazing Intensity and Ecosystem Processes in a Northern Mixed-Grass Prairie, USA." *Ecological Applications* 8:469–479.

DiFonzo, C., and R. Hines. 2001. *Soybean Aphid in Michigan*. Michigan State University Extension Bulletin E-2748. East Lansing, Mich.

Ehrlich, P. R. 1988. "The Loss of Diversity: Causes and Consequences." Pp. 21–27 in *Biodiversity*, edited by E.O. Wilson. National Academy Press, Washington, D.C.

Hatch, D. A., J. W. Bartolome, J. S. Fehmi, and D. S. Hillyard. 1999. "Effects of Burning and Grazing on a Coastal California Grassland." *Restoration Ecology* 7:376–381.

Landis, D., S. D. Wratten, and G. M. Gurr. 2000. "Habitat Manipulation to Conserve Natural Enemies of Arthropod Pests in Agriculture." *Annual Review of Entomology* 45:173–199.

Landres, P. B., P. Morgan, and F. J. Swanson. 1999. "Overview of the Use of Natural Variability Concepts in Managing Ecological Systems." *Ecological Applications* 9:1179–1188.

McNaughton, S. J. 1993. "Grasses and Grazers, Science and Management." *Ecological Applications* 3:17–20.

Michigan Natural Features Inventory (MNFI). 1997. *Natural Community Abstract for Prairie Fen*. Michigan Natural Features Inventory, Lansing, Mich.

Pyle, R. M. 2000. "Resurrection Ecology: Bring Back the Xerces Blue." *Wild Earth* 39 (fall):30–34.

Sawhill, J. C. 1990. "Facing Future Challenges to Conservation." *Nature Conservancy* 40:6–9.

Schultz, C. B., and E. Crone. 1998. "Burning Prairie to Restore Butterfly Habitat: A Modeling Approach to Management Tradeoffs for the Fender Blue." *Restoration Ecology* 6:244–252.

Schuman, G. E., J. T. Manley, R. H. Hart, W. A. Manley, and J. D. Reeder. 1999.

"Impact of Grazing Management on the Carbon and Nitrogen Balance of a Mixed-grass Rangeland." *Ecological Applications* 9:65–71.

Stein, B. A., L. S. Kutner, and J. S. Adams, editors. 2000. *Precious Heritage*. Oxford University Press, New York.

Stohlgren, T. J., L. D. Schell, and B. Vanden Heuvel. 1999. "How Grazing and Soil Quality Affect Native and Exotic Plant Diversity in Rocky Mountain Grasslands." *Ecological Applications* 9:45–64.

Vosick, D., and K. Cash. 1996. *The Role of Agriculture in Protecting Biological Diversity*. Working Paper for Seminar on Environmental Benefits from Agriculture: Issues and Policies. The Organization for Economic Cooperation and Development.

WallisDeVries, M. F. 1998. "Large Herbivores as Key Factors for Nature Conservation." Pp. 1–20 in *Grazing and Conservation Management*, edited by M. F. WallisDeVries, J. P. Bakker, and S. E. Van Wieren. Kluwer Academic Publishers, Dordrecht.

World Wildlife Fund (WWF). 2000. *Freshwater Ecoregions of North America: A Conservation Assessment*, edited by R. A. Abell, D. M. Olson, E. Dinerstein, P. T. Hurley, J. T. Diggs, W. Eichbaum, S. Walters, W. Wettengel, T. Allnutt, C. J. Loucks, and P. Hedao. Island Press, Washington, D.C.

Chapter 13

Integrating Wetland Habitat with Agriculture

Carol Shennan and Collin A. Bode

Mount Shasta, with its 14,000-foot, snow-covered peak, forms a spectacular backdrop to the volcanic hills, wetlands, and wide expanses of barley and potato fields that make up the landscape of the Tule Lake basin. Nestled in a high mountain valley in northeastern California, this area is the setting for a continuing search for creative ways to balance needs of wetland conservation with those of generations of farm families who came to the area as homesteaders in the early to mid-1900s. The stark beauty of the area is matched by a rich history reflected in petroglyphs carved into cliff faces by unknown Native Americans, in remains of the World War II Japanese internment camp, and in the faces of the people who now live and work in this sometimes harsh but beautiful land. For the past fifteen years, refuge officials, community leaders, and other groups in the area have struggled to reverse declines in the numbers of waterfowl using the Tule Lake National Wildlife Refuge without devastating the local farm based economy. Successive lawsuits and counter suits have polarized those groups with a stake in the region. The severe drought of 2000–2001 led to water supplies being completely cut off to most areas of farmland in the Klamath Basin, as well as to the Klamath Basin National Wildlife Refuge system. This move to protect endangered sucker fish in Upper Klamath Lake and threatened coho salmon downstream in the Klamath river has generated nationwide attention and intense reactions from all sides of the issues. Yet in the midst of these tensions a collaborative project has emerged to test an innovative management system that has the potential to benefit both wildlife and agriculture. This project forms the main focus of this chapter, but to understand it we need to consider the history leading up to the creation of the project.

A History of the Area

Before European settlement, Native Americans were drawn to the basin by the wetlands and lakes created from waters released as snow melted in the surrounding mountains. The size of the lake and wetlands would fluctuate yearly. In 1890, Tule Lake covered more than 100,000 acres, whereas in 1846 it shrank to only 53,000 acres (Abney 1964). Evidence from high-water marks on surrounding cliffs show that lake level variations were even greater prior to the 1800s. This variation in water level led to a rich diversity of upland, wetland, and open water habitats and a highly productive wetland ecosystem. These wetlands were home to an incredible diversity of plants and animals, including millions of geese and ducks migrating each fall and spring along the Pacific Flyway. The large concentrations of birds and small mammals also made the region home to the largest overwintering population of bald eagles in the lower forty-eight states. A glimpse of what the area was like can be gleaned from a passage in a 1959 edition of *Sports Illustrated*:

> In a mid-October day of almost any year when the southern migration along the Pacific Flyway is in mid-flight, a visitor to northern California's Tule Lake may still see a sight as full of wonder as that of the buffalo . . . the sight of some six million ducks and geese gathered in a single rendezvous. (Anonymous 1959)

The same processes that created the productive wetlands in the flat valley bottom also created deep rich soils ideally suited for farming after drainage. Between 1922 and 1948, forty-four thousand acres of the flat valley bottom lands were homesteaded and drained for irrigated agriculture (Pafford 1971). In the midst of homesteading and reclamation, Tule Lake National Wildlife Refuge was created by federal executive order, eventually encompassing 30,000 acres of the basin. The refuge was to be managed for waterfowl, but since it was superimposed on lands already ceded to the United States government for reclamation purposes, drainage for farmland continued within the refuge boundaries. As the area in wetlands continued to decline, conflicts ensued between conservation, hunting, and farming interests over the future use of the refuge.

Creation of a Dual-Function Refuge

The issue was resolved by passage in 1964 of the Kuchel Act. This Act stated that lands within the Tule Lake National Wildlife Refuge (as well as other nearby refuges) were now *dedicated* to the major purpose of waterfowl management, but *with full consideration to* optimum agricultural use that is consistent therewith and that the refuge would not be opened to homestead entry. It required that the remaining area in wetland be sustained and not reduced by further drainage. The Act also allowed leasing land within the refuge for farming that is "consistent with waterfowl management" (Sorenson and Schwarzbach 1991). This created a unique "dual-function" wildlife refuge unlike any in the country (see Fig. 13-1).

In creating a dual-function refuge, the Act recognized both the importance of agriculture and wetlands to the region's economy and the role crops played in feeding the millions of waterfowl migrating along the Pacific Flyway. Hunting, bird watching, and related industries such as bird plucking and taxidermy also provided considerable income to the basin at the time. As Secretary of Interior Stewart Udall stated during the congressional hearings in 1964:

> This refuge of more than 37,000 acres is set in the midst of rich grainlands that were once the bed of Tule Lake. The 13,000-acre sump filled with aquatic food plants, and the surrounding fertile grainfields, provide ideal habitat for waterfowl. Here the photographer can find flocks of ducks and geese that darken the sky—for millions [of] pintails and hundreds of thousands of mallard and geese gather in the refuge during the fall migration.

Today, crops produced on lands leased to farmers within the refuge are valued at around $11 million per year (1995 data) with an additional $1 million per year generated in the local community from ecotourism by an estimated 200,000 visitors. The pattern of 13,000 acres of wetlands bordered by irrigated agriculture has changed little since the 1950s, with crop production consisting of approximately two-thirds small grains (barley, wheat, and oats) and one-third row crops, primarily potato, with some onions and sugarbeet. Land leased for farming in the refuge is critical to local farmers and rural communities surrounding the refuge. Agricultural enterprises in the region are typically family-owned operations, and in many cases the refuge lease-land parcels are the most productive land to which farmers have access. Especially for the younger farmers with limited capital and land, refuge lands can make the difference between profit and loss.

The Current State of the Refuge

Up until the 1970s, the Tule Lake National Wildlife Refuge was considered one of the most important waterfowl refuges in North America. Today, waterfowl numbers have declined to less than one-third of the 1970 numbers, raising major concerns about the health of the refuge and the impact of agricul-

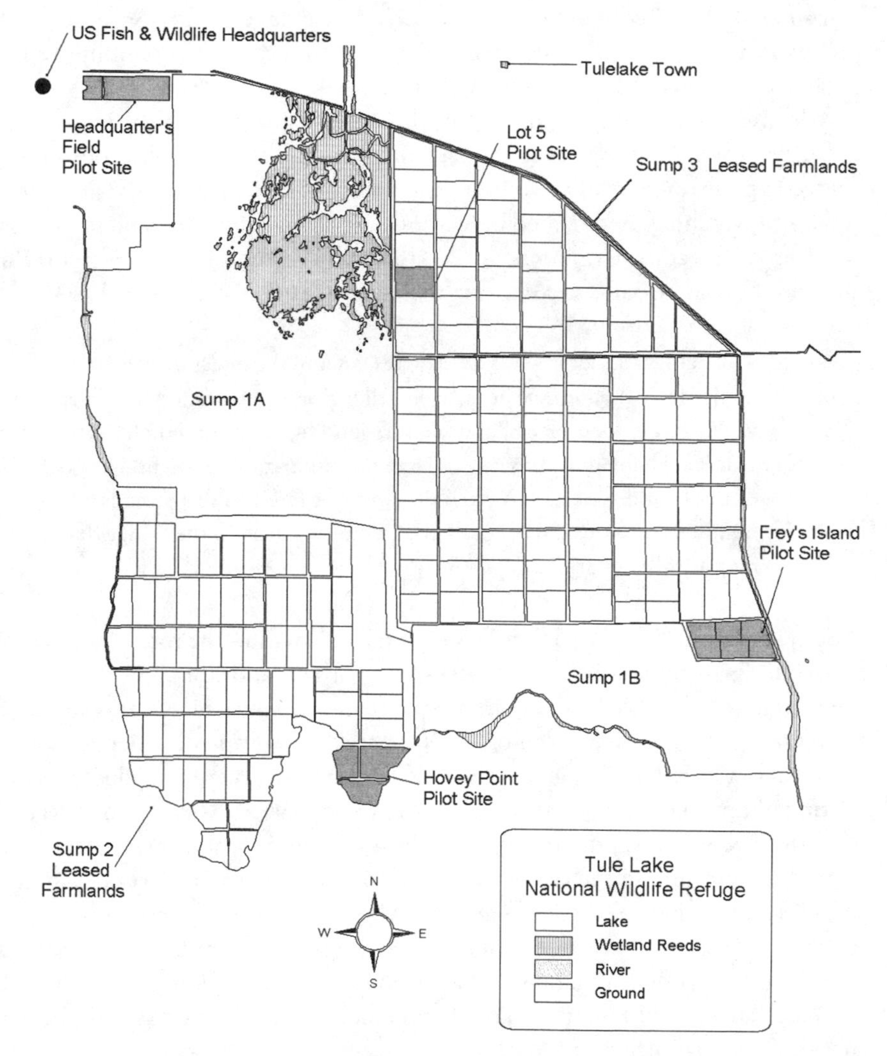

Fig. 13-1 Map of Tule Lake National Wildlife Refuge showing areas of permanent wetland, farmed areas, and pilot studies of wetland/cropland rotation.

ture that developed in the basin throughout the twentieth century. Is the use of agricultural pesticides and fertilizers leading to deteriorating water quality and contamination of the refuge, or do the declines reflect reduced numbers of birds migrating along the Pacific Flyway due to habitat loss and problems elsewhere? This question has intensely occupied many people from researchers to farmers, hunters to environmentalists, and not least the refuge managers, for the past fifteen years as they search for solutions that meet both the needs of waterfowl conservation and the local agricultural economy. In the remainder of the chapter, we will describe some of these efforts to make productive agriculture coexist with healthy, diverse wetlands.

Tule Lake National Wildlife Refuge is part of the Klamath Basin National Wildlife Refuge located in the Upper Klamath Basin of southern Oregon and northeastern California. The refuge is often the first resting area waterfowl migrating along the Pacific Flyway reach after leaving their northern nesting areas. In the fall, crops left in the fields on Tule Lake National Wildlife Refuge are an important food resource for a host of species including snow, Ross, and white-fronted geese, mallards, pintails, shovelers, and widgeon. Grain stubble is left undisturbed over the winter in parts of the refuge to provide a feed source for migrating birds, which also feed on potato crop residues remaining after harvest. In the spring, invertebrate populations that build up in the wetlands provide the major food supply for the northward migration, but volunteer grain germinating in the spring provides an additional food source. Although managed primarily for waterfowl, a multitude of other birds such as ibis, herons, egrets, bitterns, rails, shorebirds, and terns also depend upon Tule Lake National Wildlife Refuge. High populations of fish in the lake have historically attracted grebes, white pelicans, and cormorants, and during the spring and summer waterfowl, grebes, herons, cormorants, rails, and bitterns nest and raise young on the wetlands of Tule Lake. Abundant aquatic plants, fish, and invertebrates are critical to the success of these birds in nesting and successfully rearing young. In addition, the greater Klamath Basin is a major site for overwintering golden and bald eagles, with populations reaching over one thousand for bald eagles. Tule Lake also contains adult life stages of two endangered endemic fish species, the shortnose sucker (*Chasmistes brevirostris*) and the Lost River sucker (*Deltistes luxatus*), although they are not known to breed in the lake.

Water levels within the refuge are manipulated as part of the regional water storage and flood control system run by the U.S. Bureau of Reclamation. The wetlands within the refuge are constrained to the 13,000 acres stipulated by the Kuchel Act by steep-walled dikes. The lake section of the refuge is generally drawn down by 1 foot during the winter to provide storage capacity for runoff resulting from heavy storms and snow melt in the spring. During high flood conditions the current wetland and open water areas (i.e., the "Lake" or Sumps 1a and 1b) would be filled to capacity first, and then crop land (Sumps 2, 3) would be flooded if necessary under emergency conditions.

The Tule Lake refuge is clearly a highly managed ecosystem and has been since the early 1900s. What elements of this management have led to deteriorating wildlife values in the refuge? While it is possible that waterfowl declines could be related to the general decline found in Pacific Flyway populations, the adjacent Lower Klamath National Wildlife Refuge populations have remained stable or increased. Initial questions focused on pesticide contamination in the wetlands and/or reduced water quality due to eutrophication as a result of fertilizer loss from croplands. Indeed, agriculture is more input intensive in Tule Lake National Wildlife Refuge than in Lower Klamath National Wildlife Refuge due to the presence of row crops such as potatoes. Also, canals collecting drainage from cropland in Tule Lake National Wildlife Refuge discharge directly into the wetlands (Sorenson and Schwarzbach 1991).

Historically, serious impacts on wildlife from the use of pesticides have been documented (USFWS 1998). Use of persistent organochlorine pesticides (DDT, endrin, toxaphene, and dieldrin) between 1946 and the early 1980s caused death and reproductive failure in a number of species. Following the ban on use of organochlorines, levels of residues in birds and fish are now below detection limits. During the 1980s, studies were undertaken to identify more-current pesticide problems in the refuge, but no clear evidence emerged to demonstrate a causal link between species diversity and pesticide applications (Sorenson and Schwarzbach 1991). Concerns about chronic effects (such as inhibition of the enzyme acetylcholinesterase in birds that can lead to lethargy, poor motor control, nest inattentiveness, and, in severe cases, even death) from organophosphates and carbamates have led to recent banning of one insecticide product in the refuge and restriction on others (USFWS 1998). Use of pesticides within the refuge remains a volatile political issue and a primary focus of environmentalists' efforts to change or eliminate farming from the refuge.

Work has also been underway for a number of years to determine to what extent fertilizer from croplands is contributing to poor water quality in the lake and wetlands. Tule Lake has apparently always been eutrophic (high in nutrients), with high pH, low dissolved oxygen, and periodic high concentrations of unionized ammonia that can be toxic to aquatic organisms. This is similar to water quality conditions in Upper Klamath Lake, Tule Lake's primary water source. However, the water coming into the Tule Lake wetlands has also cycled a number of times through farmland, applied as irrigation and removed as subsurface drainage. Water quality fluctuates considerably, making it very difficult to determine the relative importance of natural and human-induced processes that cause periods of damaging hypereutrophic conditions.

Fish endemic to the region are adapted to eutrophic conditions and can thrive in fairly poor water conditions, but consistently high water temperatures, low dissolved oxygen, and high pH will kill them. Relatively small improvements in water quality can shift the water from uninhabitable to habitable. Efforts to improve water quality should benefit the refuge, but the general feeling is that water quality is not the primary reason for declining waterfowl numbers.

The most prominent opinion among refuge managers is that loss of habitat diversity and quality is behind the bird population declines. The Kuchel Act permanently established a fixed area of wetland and cropland, thus eliminating the wide fluctuations in water levels and marsh areas that characterized the area prior to 1905. Marshes are dynamic systems that, to remain highly diverse and productive, must experience periodic disturbances such as drying, floods, and fires. The fixed area of wetlands and stabilized water levels eliminated most of these disturbances.

Particularly important for migratory waterfowl are seasonal marshes, which are flooded from fall to early summer, then slowly dry down as summer progresses. Seed-bearing plants such as red goosefoot (*Chenopodium chenopodiodes*) and smartweed (*Polygonum lapathifolium*) germinate and grow in the moist soil that develops in these areas. By the 1990s essentially no areas of seasonal marsh existed in the Tule Lake National Wildlife Refuge. Furthermore, the tule (*Scirpus acutus*) marshes had shrunk from 7,000 acres in 1956 to 3,000 acres in the 1990s, probably due to the stabilized water levels precluding the periods of drying down that are needed to stimulate new establishment. The remainder of the 13,000 acres designated as wetlands now consists of wide expanses of shallow open water with no emergent vegetation, which provide limited habitat and food for waterfowl.

The other functions of the refuge—namely water storage and agriculture—are also experiencing problems. Between 1958 and 1986 sedimentation reduced the average water depth in the lake by about 14 inches, representing a loss of 12,800 acre-feet of storage and flood-control capacity. Sedimentation results from airborne deposition of soil particles generated by wind erosion of bare fields and from sediment carried in rivers entering the wetlands. Sedimentation reduces the amount of deep-water habitat available for the endangered sucker fish. It also reduces depth in emergent marsh areas and has nearly eliminated the marsh as nesting habitat for diving ducks and colonial nesting waterbirds.

Agriculture is also experiencing a gradual reduction in soil quality within the refuge lands. After fifty years of farming, the originally high organic matter content of the reclaimed soil has declined. As a result, the capacity of these soils to provide nutrients has also declined. In addition, chronic problems with soil-borne pests such as Columbia root knot nematode (*Meloidogyne chitwoodii*) on potato, and fungal diseases such as white mold of potato and white rot of onion threaten crop production. Growers have maintained productivity by increasing inputs (fertilizer, pesticides), which increase their production costs and intensify concerns about contamination of the wetlands.

Wetland/Cropland Rotational Management

Although the problems facing the refuge seem daunting, one idea had been raised by many of the actors in this unfolding drama over the years. These included the U.S. Bureau of Reclamation, a local grower group called the Klamath Basin Water Users Protective Association (1993) and most recently the U.S. Fish and Wildlife Service staff at the Tule Lake National Wildlife Refuge. The idea was to re-flood some cropland areas to create new wetlands with a diversity of flood regimes (i.e., seasonal and permanent), then subsequently drain other areas of existing wetlands with low wildlife values to create new farmland of higher organic matter and free from crop pathogens. This strategy of rotating wetlands and cropland could potentially improve habitat diversity and reduce the need for pesticides while improving the productivity of croplands. (See chapter 12 for an example of wetland/cropland rotation in Texas rice fields.) A second strategy, conversion of low-lying areas currently in cropland to permanent wetland, would increase deep-water habitat for fish and improve overall water storage capacity. These deep-water areas would be maintained over the long term and not placed in rotational management.

This wetland/cropland rotation approach had the potential to benefit wildlife, agriculture, and rural community interests in the area. Crop production cycles would function as a disturbance to break wetland succession and enable juvenile marsh stages to be established upon re-flooding. Subsequently water management could be varied to create the desired mix of habitats. For example, flooding some areas seasonally and allowing them to dry down over the summer would simulate marsh edge habitat while others could remain flooded year round. Thus, a mosaic landscape with units at different stages of marsh succession interspersed with areas in crop production would be created. Rotation out of crop production would provide a break in key pest cycles. In controlled experiments, flooding has been shown to reduce propagule levels of the fungi causing white rot and white mold. Increased organic matter inputs from decomposed wetland vegetation should also increase soil fertility and improve soil physical properties. This newly created cropland provides an excellent opportunity to test reduced-pesticide cropping strategies on land initially free of serious pathogen and nematode problems.

Pilot Testing

Beginning in 1995, two types of rotation cycles have been investigated: a short-cycle (three to four years in both wetland and cropland phases) and long cycle where the rotation length will be on the order of fifteen to thirty years. The rationale for the short-cycle system is to create early stages of seasonal marsh succession in the wetland phase dominated by readily decomposable annuals followed by drainage before significant perennial vegetation becomes established. The absence of woody material should make it easier for farmers to transition into crop production but still provide the pest control benefits of flooding. The long-cycle system would create mature and diverse marsh habitat with areas of established perennial emergent vegetation. In this case, the transition back into farming would require that the farmer burn or cut down and then incorporate large amounts of biomass into the soil before planting. Crop yields would likely be reduced the first year coming out of mature wetlands because of problems with seedbed quality and regrowth of "wetland weeds" such as tules and cattails. However, this system provides for significant periods of continuous cropping essential to the farm economy, and with the high organic matter inputs into the soil during the long wetland phase, soil quality and fertility should resemble that of the original virgin soils.

To test how well this system could work, a team of researchers, refuge managers, cooperating farmers, the U.S. Bureau of Reclamation, and the local irrigation district established a series of pilot test sites, each between 40 and 140 acres (Fig. 13-1). Three sites converted cropland into long-cycle wetland and four into the short-cycle rotation. During the cropping phase of the rotations, land was leased and managed by local farmers in accordance with agreed-upon project guidelines. This ensured that results reflected realistic expectations for commercial farming. During the wetland phase, the sites were managed by staff at the Tule Lake National Wildlife Refuge. Management records and a variety of monitoring data have been collected for each site, and the findings so far are very promising.

Wetland Vegetation

Results from the first five years of the pilot projects have been very encouraging. Desirable seasonal wetland vegetation has developed within two to three years in sites previously farmed for decades. Important wetland species became abundant within one to three years at all sites, in particular the seed producing annual plants, red goosefoot and smartweed. Further, noxious weedy species that appeared after the first year of seasonal flooding were effectively eliminated by the second year. The most rapid development of wetland vegetation occurred close to established wetlands, which appeared to serve as sources of seeds. Analysis of seeds present in the soil prior to beginning the wetland management showed that after forty to fifty years of farming only about 15 percent of viable seeds in the soil "seed bank" were wetland species, further emphasizing the importance of proximity to wetland seed sources.

Bird Use

Bird usage of the pilot sites has also been encouraging. According to surveys conducted by Tule Lake refuge staff during 1998 and 1999 fall migrations, between 19 and 35 percent of the ducks and geese present in the refuge were found in the flooded pilot sites that represented only 4 percent of the total wetland area. Similarly in May 1998 around 23 percent of duck breeding pairs were found in the pilot sites, although the percentage varied widely from species to species. Bonsignore (1998) learned that seasonal wetlands were particularly important as feeding and loafing sites for dabbling ducks such as northern pintail, American widgeon, mallard, and gadwall but less so for diving ducks such as bufflehead and lesser scaup. Seeds produced by the seasonal marsh plants provided an important food source for dabbling ducks in the fall migration, whereas in the spring migration abundant aquatic invertebrates found in the newly created marshes were the main food source. In contrast, geese and swans depended heavily upon surrounding agricultural fields for their food (leftover grain, potato, or green browse from young cereals) but they used the pilot sites extensively for loafing.

Changes in Soil Quality—Fertility and Nematodes

For the rotational management scheme to work for farmers as well as wildlife, wetland cycles should improve agricultural productivity through enhanced soil fertility and/or pest control. So far, two sites have been returned to crop production after completing two to three years in wetland management. In both locations, farmers were able to achieve excellent barley and potato yields, even in sections of the fields where they greatly reduced their nitrogen fertilizer use. Soil samples were collected from the pilot sites before they entered the wetland rotation and re-sampled in subsequent years. Levels of available phosphorus have increased by 40 to 70 percent after three years of wetland management in three of the four sites analyzed. Also, at one site total soil nitrogen increased by about 16 percent. This suggests that farmers can transition back into crop production after short wetland cycles without incurring yield losses.

The potential for wetland cycles to reduce levels of the Columbia root knot nematode, a major pest of potatoes, was tested at one pilot site with a known history of serious nematode problems. The high levels (up to 3,500 larvae per liter of soil) observed across much of the site prior to flooding were completely eliminated after the first year of seasonal marsh management. Re-sampling during the first cropping season also failed to find detectable levels of these nematodes, providing further evidence that the short wetland cycle had effectively reduced nematode populations.

Water Quality

Do wetland cycles have the potential to improve water quality? The answer to this question is not straightforward, since water quality varies greatly from season to season, year to year, and location to location. Nonetheless, a few general statements can be made. Seasonal wetlands generally maintained good water quality in terms of low nitrate-N, ammonium-N, and orthophosphate-P in the surface water, although occasional spikes of increased nutrient concentrations were observed. Dissolved oxygen generally fell each spring, but the timing of the decrease appeared to be related to temperature: the warmer the spring, the earlier the decline in dissolved oxygen. When water levels were drawn down in June to establish the desired annual moist soil vegetation, movement of water through the soil profile resulted in lower dissolved oxygen and higher orthophosphate-P and ammonium-N in subsurface drainage water. Fortunately, pH and temperature also tended to be lower in the drainage water, which would reduce the amount of highly toxic un-ionized ammonia. Thus, net water-quality impacts of seasonal marshes on the refuge will depend on timing of release of water during the draw-down, the balance of surface and subsurface drainage, and the volume flows and background water quality of the drainage canals or wetlands they drain into (Shennan unpublished).

The Future of Wetland/Cropland Rotations in Refuge Management?

Is it likely that this innovative approach to sustaining agriculture and productive wetlands within the same landscape will become part of the management of the Tule Lake National Wildlife Refuge? It is hard to say. Some elements of the strategy, such as use of short flooding cycles to reduce soil pest populations, have generated considerable interest among farmers and refuge managers. For instance, in the fall of 1999 two adjoining lease units with a history of nematode and white rot problems were put into seasonal wetland management to create additional seasonal marsh habitat and improve the land for future crop production. But, whether refuge management overall will shift to an integrated system of wetland/cropland rotations and deep-water habitat will depend not only the economic feasibility and compliance questions to be addressed at the refuge level, but also on how some larger issues are resolved.

One major issue is water. Regional conflicts over water have intensified, especially during drought years. A second issue is the long-term viability of agriculture in the region. Crop prices have declined steadily, and serious barriers exist for exploring new crop markets. Third, it has proven difficult to change the widely held perception that conservation and farming should remain separated in the landscape. Finally, there is no mechanism for reinvesting the refuge's income from agricultural leases directly into wetland management. Any one of these issues could derail movement toward an integrated wetland/cropland management for the refuge.

The first issue, intensified conflicts over water allocation, illustrates the difficulty of effectively balancing competing regional needs. Water availability fluctuates widely from year to year, and in many years is insufficient to meet all the competing needs in the watershed. Historically, agricultural irrigation needs were given highest priority, but the situation has changed dramatically in the last decade. First, the shortnose and Lost River suckers were listed under the Endangered Species Act and as a result new lake-level requirements were put in place for Upper Klamath Lake and the current open water areas in Tule Lake. Subsequently, coho salmon downstream in the Klamath River were listed as being threatened, leading to increased projected demand for downstream river flows at key times of the year. These listings, together with a heightened awareness of tribal trust obligations to the Klamath Basin Tribes, led the Department of Interior to recognize the Klamath Basin Tribes and requirements under the Endangered Species Act as having senior water rights over either agriculture or refuge needs for waterfowl habitat (USFWS 2001). This means less water for refuge wetlands and agriculture, and heightened conflicts between farmers and the refuge. Indeed, water-demand models predict that in half of future years, 60 to 75 percent of refuge wetlands would be dry at peak fall waterfowl migration.

In response to these predictions, the U.S. Fish and Wildlife Service has reevaluated the compatibility of farming within the refuge, now stating that "the refuge farming program is compatible and consistent with the primary purpose for which the refuge was established . . . only if sufficient water is available to maintain refuge wetlands first, followed secondarily by water use on agricultural crops" (USFWS 2001). In practice, water deliveries to farms could be halted in mid-season to protect waterfowl in the refuge. When this was proposed in 1999, the Tule Lake Irrigation District filed a lawsuit against the proposed changes citing disruption of local economies, loss of income, loss of wildlife habitat, increased depredation of crops in private lands by waterfowl, and increased noxious weed abundance.

The water situation is yet to be resolved. The U.S. Bureau of Reclamation is still developing an integrated water management plan for the Klamath project, and the U.S. Fish and Wildlife Service is revising an environmental assessment, Implementation of the Agricultural Program on Tule Lake National Wildlife Refuge (USFWS 2001), in response to the Tule Lake Irrigation District agreeing to drop the lawsuit. Both processes require the agencies to grapple with balancing many competing legal and moral perspectives. For example, the needs of individual species are in competition (salmon versus suckers), as well as different habitats (open water and rivers for fish versus wetlands for waterfowl), and different sections of society (tribal groups, environmentalists, hunters, fishers, farmers, etc.). Clearly, the potential for a local solution to ecosystem management in Tule Lake National Wildlife Refuge will depend heavily on how the larger regional water issues are addressed. This process was dealt a severe blow when a record dry winter in 2000–2001 led the U.S. Bureau of Reclamation to cut off all irrigation water supplies for the summer of 2001 in order to meet endangered fish and tribal fisheries requirements.

This move brought conflicts around endangered species and local economies to center stage. Indeed, Senator Herger (R-Calif.) was quoted in Klamath Falls' *Herald and News* as calling the Klamath Basin the "poster child" of the anti-ESA movement. A variety of efforts are being made to deal with the short-term crisis, including limited emergency federal and state aid to farmers and local community programs. Yet these efforts did not prevent law suits from being filed by individuals and on behalf of the agricultural community. The wildlife refuges have also suffered from lack of water. Recent agreements have enabled both the refuges to purchase water from local irrigators to meet the fall migration water requirements and the California Waterfowl Association to purchase the limited grain crop in the basin to provide waterfowl food. Clearly, long-term solutions are needed, since severe droughts will undoubtedly recur. Avenues being pursued include assessing groundwater supplies in the region that could be used in drought years, looking at buy-out schemes to enable farmers to take land out of production for wildlife habitat, and re-examining restoration and irrigation options for the basin. In addition, a National Academy of Sciences panel is reviewing the data upon which the lake-level and stream-flow requirements under the ESA are based. Whatever this review's outcome, any future plans for the Klamath watershed will have to address how to balance the needs of fisheries, waterfowl, different tribal groups, and agriculture in a whole, watershed-fashion. Success or failure will have serious repercussions for agriculture and environment conflicts nationwide.

The second major issue is the long-term viability of agriculture in the region. Farmers are facing serious economic problems driven in large part by the decline in prices for the major crops grown in the basin. Prices in the 1990s for barley and potatoes, two major crops in the region, averaged 45 percent of what they were in the 1960s and 62 percent of what they were as late as the 1980s. Many growers and the University of California Extension personnel are looking for alternate crops that may offer greater economic returns, but the long distance to markets or processing plants makes it difficult for them to compete with other regions. Although integrated wetland/cropland management may have the potential to enhance the local economy in the future through reduced farming costs and increased income from tourism and hunting, many farmers are facing bankruptcy now or are deciding to leave farming for other work. This discourages farmers from shifting to new management systems that may involve any increased economic risk.

Finally, for the integrated wetland/cropland management ideas presented here to be implemented, it will require a significant shift in how agencies and other groups have conceived the relationship between conservation and economic activity. Traditionally in the United States, wildlife refuges, national parks, and other conservation areas have largely been created to be exclusive of economic activity in order to sustain their "wilderness" or "wildness" values. Policies, expectations of visitors, and positions of many environmental groups have thus been built around separating these functions in the landscape. The unique dual-function character of the Tule Lake National Wildlife Refuge since the passage of the Kuchel Act has been at odds with this vision. It has proved difficult at times for managers to implement national policies that have been developed for refuges that do not have significant intensive economic activity within their boundaries. A good illustration of this is the issue of pesticide use on the refuge. During the recent development of the integrated pest management plan, refuge staff had to balance the needs of farmers with national U.S. Fish and Wildlife Service policy of eliminating or restricting pesticides and herbicides within refuges.

So far there has been a spectrum of reaction to the concept of intensive farming interlinked in a landscape mosaic with wetlands. Some groups such as Ducks Unlimited, the California Waterfowl Association, and The Nature Conservancy have supported the pilot studies, whereas others are waiting for specific management plans to respond to, and others still question whether farming has a place in a national wildlife refuge.

Part of the beauty of the integrated wetland/cropland management is that it provides a mechanism for mutual support of both economic production and conservation. It could give the local economy a vested interest in the refuge.

In theory, if the appropriate institutional mechanisms were developed, revenues from agricultural leases in the refuge could be used to improve and maintain the wetlands and other wildlife habitat. (In 1996, lease revenues were $1.9 million, larger than the refuge operating budget.) Currently, some of these monies do return to the region in the form of payments to local government and the Tule Lake Irrigation District, but a strong link to support conservation needs in the refuge has been lacking. There might be a greater interest among environmental groups to support integrated wetland/cropland management if some revenues from agriculture went directly to support habitat management goals.

In 1999, a working group was formed made up of individuals from the refuge, the Tule Lake Irrigation District, University of California Intermountain Research and Extension Center, the Klamath Office of the U.S. Bureau of Reclamation, and the California Waterfowl Association. This group produced a proposal examining four alternatives for "Integrated Land Management" in the Tule Lake National Wildlife Refuge that has been submitted to the U.S. Fish and Wildlife Service. Time will tell whether the wetland/cropland management approaches described here will be implemented, and in what form. What has been demonstrated, however, is the potential for sustaining productive wetlands and cropland within a landscape, and a recognition that we don't have to restrict our ideas for habitat and wildlife conservation to areas devoid of intensive farming. This lesson has relevance for how we as a society struggle to balance the needs of humans and nature.

Acknowledgment

We would like to acknowledge the numerous and invaluable discussions, advice, and assistance provided by the USFWS staff at the Tule Lake NWR, the UC Intermountain Research and Extension Center, Tule Lake (IREC), U.S. Bureau of Reclamation, Klamath Falls, many growers in the Tule Lake Basin, and others concerned about the basin. In particular, we single out Harry Carlson (IREC) and Dave Mauser (USFWS), for their tremendous efforts to create workable solutions.

References

Abney, R. M., 1964. "A Comparative Study of Past and Present Conditions of Tule Lake: A Report Submitted to the Bureau of Sport Fisheries and Wildlife." *Tule Lake National Wildlife Refuge Report on the Request of U.S. Congress*. U.S. Fish and Wildlife Service, Tulelake, Calif.

Anonymous. 1959. "Is Tule Too Good for Ducks?" *Sports Illustrated* 11:37.

Bonsignore, C. M. 1998. "Aquatic Bird Use of Seasonal Wetlands Created by Flooding Agricultural Fields on the Tule Lake National Wildlife Refuge." Master's thesis, University of Washington, Seattle.

Klamath Basin Water Users Protection Association. 1993. *Initial Ecosystem Restoration Plan for the Upper Klamath Basin.* Klamath Basin Water Users Protection Association, Klamath Falls, Ore.

Pafford, R. J., Jr. 1971. *History and Future of the Klamath Reclamation Project.* Bureau of Reclamation, Region 2, Klamath Falls, Ore.

Sorenson, S. K., and S. E. Schwarzbach. 1991. *Reconnaissance Investigation of Water Quality, Bottom Sediment, and Biota Associated with Irrigation Drainage in the Klamath Basin, California, Oregon 1988–89.* Water Resources Investigations Report No. 90-4203. U.S. Geological Survey, U.S. Fish and Wildlife Service, and U.S. Bureau of Reclamation, Denver.

U.S. Fish and Wildlife Service (USFWS). 1998. Klamath Basin Refuge Complex Integrated Pest Management Plan. Klamath Falls, Ore.

———. 2001. Implementation of the Agricultural Program on Tule Lake NWR. Tulelake, Calif.

Part IV

Steps Toward Agroecological Restoration

What steps can we take to bring about agroecological restoration? How do we end the extreme specialization in crops and livestock, this forced simplification of the landscape that destroys natural habitats and ecosystem services? How can we keep from merely accepting the "inevitable" and turning our eyes and minds away from such an intractable problem? It would be so much easier to stay focused on protecting those lands already set aside in parks and wilderness, and fighting for more public parks, forests, and nature preserves outside agricultural areas. But when we consider that 50 percent of the United States is cropland, pastureland, and rangeland owned and managed by farmers and ranchers (NRCS 1996), we realize that even protected lands cannot escape the negative impacts of industrialized agriculture. We must influence all these land managers to farm as if nature mattered, to value biodiversity and refrain from practices that destroy natural habitats.

A knee-jerk reaction is to put all the responsibility on farmers and tell them to shape up. But we can't just preach at farmers; the problems belong to all of us who eat and vote. Farmers need help from conservationists, consumers, and government policymakers to release them from the powerful hold of trends in industrial agriculture. They will need nontraditional advice and tools to make transitions that succeed economically for them while benefiting the land and society. They will need cooperative processes with neighboring farmers and community leaders for marketing and for conservation and nature restoration across property lines. Consumers must pay attention to where their food comes from and care how it was grown and how the land was managed so they can support those good stewards of the land through their purchases. And a complete overhaul of federal farm policy must occur, first to stop funding highly destructive agricultural practices for the sake of high yields so grain handlers and processors can buy cheap commodities. Then, government programs must provide incentives and rewards for those farmers that produce not only commodities, but also other environmental and social benefits for society. The chapters in Part IV discuss the steps needed to move toward agroecological restoration.

Rhonda Janke, agronomist at Kansas State University, explains in chapter 14, "Composing a Landscape," the nitty-gritty decisions farmers must make to change their farming systems. She describes a useful process called *whole farm planning*, which helps farmers design their production practices to take protection of natural resources into consideration while still satisfying family and financial goals. Each farm is different, and the particulars of land management sometimes make decisions by the farmer quite complicated, as her story about a Kansas dairy farm shows. But whole farm planning and assistance from nonprofit organizations and empathetic U.S. Department of Agriculture extension experts at land grant colleges of agriculture can enable them to see options and create solutions that fit their particular operations.

Chapter 15, "After the Deluge," is about working with farmers in a larger landscape context. The author is Cheryl Miller, a staff member of the National Audubon Society, who represented her organization in a lengthy flood control mediation that took place in the Red River Valley following the 1997 flood. The region's extensive tallgrass prairie, wetlands, lakes, and woodlands had almost all been converted to farmland through extensive drainage. Natural river systems were degraded by dam construction, channelization, and sedimentation. This chapter is a case study of how ecosystem management principles and processes were introduced into this agricultural setting in order to address both restoration of the area's natural features and flood control for farms and towns. This was made possible by a new approach that advocates participation by a broad array of community members throughout the entire process of designing flood control plans, rather than the usual practice of considering comments from the public after the decisions have all been made.

Chapter 16 requires the reader to step back and look within, to consider what role personal motivation plays in restoration of the natural world in farming landscapes. In "A Refined Taste in Natural Objects," writer Beth Waterhouse explores what might cause people to care for *all* of the land, to be as outraged about damaging farmland practices as any other improper land usage, such as forest clearcutting. What do we find within personal experience that reconnects people (not just farmers) emotionally to the land that could become the basis for public rejection of ecological sacrifice in farming country? This chapter reminds us that we also need emotional tools to create big changes in the way land is managed. Reasoned processes and technological breakthroughs will not be enough on their own; "we must speak about loving attachment to the earth."

Chapter 17, "Food and Biodiversity," by Dana Jackson looks at the way we eat in the United States and the changes in diet that are needed in order to return greater biological diversity to agricultural landscapes. Consumers must be motivated to shop for the "sustainable table," but where do they find the foods that have been raised sustainably? Jackson questions whether the new USDA Certified Organic label actually fits the concept of sustainable. She describes some of the regional labels and ecolabels that differentiate sustainable from conventional products and discusses the role environmentalists should play in building market demand for such foods.

George Boody, executive director of Land Stewardship Project, writes about the role of government policy in chapter 18, "Agriculture as a Public Good." He traces the failure of agricultural policy to protect the land and keep family farming economically viable. Over half the total income for farmers producing major commodities in 2000 came from subsidies paid for by taxpayers, not from bushels produced and sold. The chief beneficiaries of such farm programs are the suppliers of inputs, processors, and retailers, and not farmers or the public at large. Boody proposes a new framework for farm policy that puts public interests ahead of corporate profits and provides incentives for farms to become natural habitats.

Part IV does not lay out the entire path to agroecological restoration for us, but it suggests steps we can take to move in that direction. We know that there are enormous obstacles not discussed in this book, such as population growth, the power of international corporations, and global trade. They add weight to the inevitability argument. But we must remind ourselves that the industrialization of agriculture is not a path but a treadmill with no destination other than destruction of the land. We owe it to our children and the other organisms with whom we share the earth to start down that path toward agroecological restoration.

References

Natural Resources Conservation Service (NRCS). 1996. *America's Private Land: A Geography of Hope*. U.S. Department of Agriculture, Natural Resources Conservation Service. Washington, D.C.

Chapter 14

Composing a Landscape

Rhonda R. Janke

I am walking through the farmer's market on a bright, clear, sunny day. I wander over to a stand with a banner: "Farm Fresh Peaches." The produce looks fresh, the family looks healthy and happy, and I soon enjoy not only the taste of fresh, ripe peaches, but also of cherries, greens, and other farm-fresh goodies. My next stop is to visit a woman selling garlic. Lots of garlic. She has about two dozen baskets, each with a different variety, carefully labeled with the name of the variety and its various attributes: hot, mild, good for cooking, best in sauce, and so forth. I am enchanted and envious of her skill and her dedication to this culinary delight. I purchase the variety pack, since I can't decide which one is best. Another farmer is selling frozen beef, which he raised himself, and a new crop of potatoes. He is a real "meat and potatoes" kind of guy.

How many of us have encountered a farmer at the market or on a farm tour and thought, "I would like a life like that." It seems simple and bucolic from a distance. They go home after the market, put away the baskets, and get up the next day and grow more good food for us to eat. We don't see the countless decisions they must make each day, balancing finances with an ongoing wish list of farm supplies and improvements. We don't see the stress of needing to spend time in five areas at once, each crop demanding attention at critical points in the growing season, caring for the sick calf on the coldest day of the winter. We don't see the effort that goes into gathering information only to find it contradictory. "Leave the crop residue on the field to reduce soil erosion" versus "completely destroy all crop residue to control the overwintering insect and diseases." "No-till will conserve soil and keep it from contaminating waterways, even though it requires more herbicide," versus "when you apply herbicide, make sure it is incorporated (tilled) to prevent it from running off into the stream."

Farmers are also bombarded with advertising on their local farm radio stations and in farm production magazines, much like those of us in the nonfarming sector are told that we will be better, prettier, smarter, and richer if we buy product "x." Farming has become product, rather than process, oriented. The prices farmers receive are dropping, while the prices they pay for inputs are still rising. Neighbors are going out of business, and most families have one, if not two, off-farm jobs to help support the family. Time to farm is getting shorter all the time.

To add insult to injury, the local newspaper says that farmers are responsible for contaminating the state's streams and rivers. Air-quality problems are also caused by dust due to tillage operations and harvesting. The Farm Bureau and other organizations go to bat for the farmers, sometimes with too much enthusiasm, reassuring farmers that they are not the cause of the pollution, that it is from cities and suburbs instead.

Who should the farmer believe? If you are that peach farmer, or garlic grower, or beef producer, are you sitting in your hammock sipping lemonade, or are you at a farm meeting trying to figure out what in the heck is going on? So much for that bucolic lifestyle.

In Part I, the authors made the case for *why* nature conservation should be pursued on privately owned agricultural land. In this chapter, I present suggestions for *how* to engage the farmers in change that reflects stewardship of resources, provides protection of wild species and their habitats, and minimizes the off-farm environmental impact of farming activities. The first step toward change is a process called *whole farm planning*.

Whole Farm Planning

The goal of whole farm planning is to match personal and farm goals with the strengths and limitations of the farm resource base in the context of economic realities to come up with an integrated whole that is stronger and more resilient than the sum of its parts.

In its simplest form, whole farm planning can be distilled into four basic steps (Figure 14-1). Goal setting is step one. Goals of the entire family need to be considered, not just those of the primary farm operator. Other nonfamily members, such as close neighbors, long-term employees, and landlords, need to be included in this process, if appropriate.

Inventory of resources is step two. This step is closely linked to goal setting, and the two should be completed in tandem, since one does not make sense without the other. Resource inventory includes taking stock of natural resources such as land, water, and other physical attributes of the farm; economic resources such as working capital, credit, and equity; and human resources such as computer prowess, mechanical abilities, and experience with livestock. The goals on the farm need to fit the resources that are either available or obtainable. For example, one could not run a hunting preserve on 2 acres of land or plant asparagus on 2,000 acres of nonirrigated land in western Kansas.

Step three is to write the plan. This can take several forms and can involve the use of planning tools such as inventories, maps, crop rotation diagrams, and economic spreadsheets.

Step four is to monitor the results. One way to do this is to set benchmarks of success. For example, a certain yield per acre might be a goal, or a certain number of animals raised and sold. Income goals for each enterprise could be established. Environmental goals might include reseeding a buffer strip along a stream to reduce run-off and erosion or to establish a high-value crop in a wooded area, such as shitake mushrooms or ginseng, while preserving the next generation of healthy trees. Another way to monitor is to set up a series of "red flags"—indicators that specific decisions should be reevaluated. When a red flag is triggered, then the enterprise or action must be reconsidered immediately and not five years down the road and too late for corrective action. In fact, fine tuning and readjustment are to be expected and anticipated as part of the planning process.

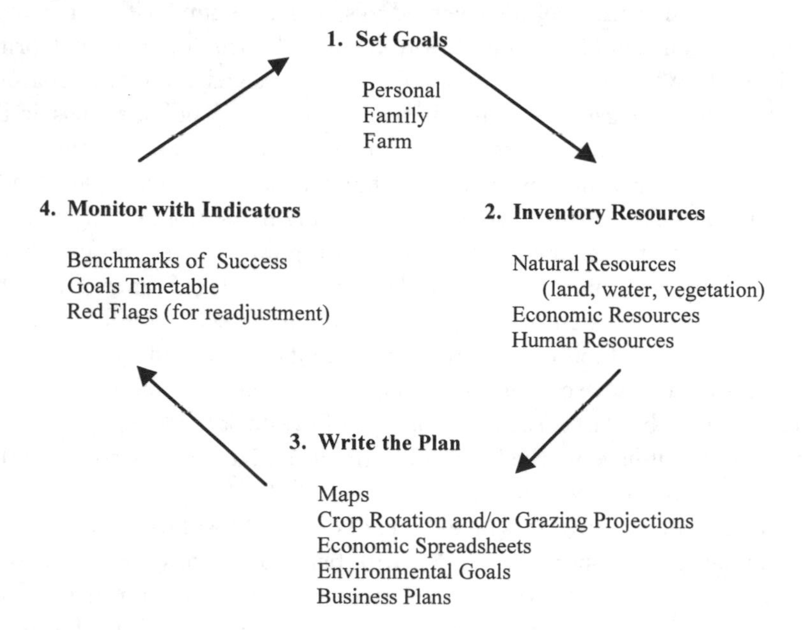

Fig. 14-1. The four basic steps used in whole farm planning, and accompanying goals, tools, and resources needed to carry them out.

A whole farm plan is not like an architect's blueprint that points to a finished product. It is more like Mary Catherine Bateson's idea of "composing a life" (1989). She writes, "Each of us has worked by improvisation, discovering the shape of our creation along the way, rather than pursuing a vision already defined."

Current Situation—Aren't Farmers Doing This Now?

We conducted a revealing set of interviews with Kansas farmers in 1995 to find out what kinds of tools farmers were using now to do farm planning and which tools they could imagine would be helpful to them in the future. We met with two groups of farmers who identified themselves as practicing or moving toward "sustainable agriculture" and two groups of farmers who were recommended as good farmers by their county agricultural extension agents but who would not have identified themselves with the word "sustainable." We wanted to get a range of opinions from a broad spectrum of articulate but representative farmers.

Right away, we noticed a stark difference between the two groups in their goals. The sustainable agriculture farmers often listed multiple goals, including inclusion of family members in farm activities, as well as financial and environmental goals. They also wanted to spend more time with family members, on personal hobbies, or on involvement with their community (Norman et al. 2000). When the other group of farmers was asked about farm goals, the first response was generally a blank look. When asked again, some said that they hoped to stay in business, or to "keep doing what they are doing."

Another surprising observation is that few farmers have a clear understanding of which parts of the farm are making them money and which are financial black holes. While the media, farm publications, and universities often encourage farmers to focus on the bottom line, few farmers know their break-even cost for a bushel of grain or a pound of beef. An informal survey I conducted by talking to several county agents, farm management professionals, and farm advisors suggests that only 5 percent of farmers calculate out their return each year for each crop or type of livestock that they produce, and only 25 percent have worked with their financial data enough to know their returns on any given enterprise.

Some farm planning does take place, of course. Most farmers have an aerial photograph, or other type of map, with field boundaries, streams, and soil types noted. Some may have a soil conservation plan written for them by their local Natural Resources Conservation Service (NRCS) office. Most farms keep track of total returns and expenditures for tax purposes. Farming also requires the use of inventories, such as bushels of grain, number of livestock, though some of these figures are tracked by the grain elevator, or are kept in one's head, such as "there are sixty head of cattle grazing on the back forty (acres)." Fewer farms keep written records of crop rotations over time and do much of their labor budgeting on the fly.

Farmers who choose to sell "certified organic" products are the exception. They must complete a questionnaire about their nutrient management, soil improvement, and pest management plans. They submit farm maps with fields identified by number and a five-year rotation plan with crop and input histories for each field. Farmers who have completed these applications often feel that they benefited from the planning required for certification.

Environmental Assessment Tools

In our review of the whole farm planning literature, there were two tools that were specifically designed for environmental assessment: Farm-A-Syst, developed in the United States, and the Ontario Environmental Farm Plan (OFEC 1996). Their implicit goal is to reduce nonpoint-source pollution from agriculture in order to avoid future regulations if possible. Farmers attend two daylong workshops and complete a lengthy checklist to determine if components of the farming operation create environmental risk or create risk for family health and safety. Conditions and practices on the farm are ranked on a four-point scale from "serious problem" to "optimal." The Ontario model is the most complete, including home well construction and water testing, septic system location, fuel storage, field crops, dairy parlor waste, feedlots—the whole farm. At the end of each section, the user lists all the items with suboptimal ratings, writes down possible solutions, and creates a timetable for addressing the concerns. Throughout the notebook, lists of resources are provided, including printed material, agencies, and programs.

One of the strengths of this program is that it is voluntary and confidential. In fact, before the Ontario Environmental Farm Plan could get off the ground, the province of Ontario had to pass a special resolution stating that the information in the notebooks could not be used against the farmers in a court of law. Once the checklist is completed, it is given a number and the environmental assessment and the action plan are reviewed by a local committee of three farmers. The workshops in which farmers develop the notebooks are offered by a nongovernmental agency that has no regulatory authority. If the notebook and action plan is deemed appropriate, the farmer is eligible for technical assistance and cost-share funds.

By the sixth year of the program, over 16,000 farmers have participated in workshops, and approximately 8,000 have completed notebooks, out of a total of 63,000 farmers in Ontario (McLean 1999). This success was partly due to the broad coalition of farm organizations who participated in the design-and-review process. However, the overriding factor seemed to be the looming specter of regulation if agriculture "did not clean up its act." The cost-share provided some incentive, but the amount was so small that many farms ranked this low on the list of reasons for participation, and some completed the program without claiming their cost-share funds. In a survey conducted during the second year of the program, 73 percent of the respondents felt that the self-assessment aspect of the program was more important than the grant (Bidgood 1995).

Encouraged by Ontario's success, a similar tool has been developed for Kansas, called the "River Friendly Farm Program." Many of the questions, and the same rating system, have been used in the Kansas version. A goal-setting section has been added, since many of the items in the action plan need to be placed in the context of a particular set of farm goals. A farmer who is about to retire will deal with a poorly placed feedlot very differently than one who is young and still expanding the farm. We are also in the process of linking the River Friendly Farm assessment program to a financial planning program, since knowing one's financial situation and outlook for the future is also necessary for one to begin to address the items in the action plan.

The River Friendly Farm Program has received positive feedback, but more incentives will be needed for this to be adopted on a wide scale in Kansas. A cash payment ($250) was offered to the thirty farmers who participated in the pilot test of the notebook, and we asked them what incentives would be required for their neighbors to participate. Some responded that the cash payment would help and also opportunities to obtain cost-share from existing or newly funded programs. One person simply wrote "good luck" on his questionnaire. Endorsement from the major farm member organizations in Canada was a prerequisite for their success, and we see this as phase 2 for our program development here in Kansas.

Barriers

What are the barriers to whole farm planning? Many farmers point to a lack of time. Many prefer field work to "book work," and, in fact, many farmers' spouses, rather than the farm operator, are in charge of finances. Some will honestly respond that they choose not to write down farm goals out of fear. "What if I don't reach my goals?" Some simply don't see the need: "If it ain't broke, don't fix it."

Family dynamics play a hidden but startlingly important role. One farmer, in business with his brother, responded that if they both wrote down their farm goals, they would probably not continue farming together. One brother wanted to expand the livestock portion of the operation and the other the crops. They each continue in their respective paths, in a tense partnership, until the banker or other outside entity forces them to get back "on the same page." For other families, intergenerational issues of conflict and control arise. Some fathers don't relinquish control over decision making until their health requires it, and so the offspring do not get a chance to influence the direction of the farm until they are well into their forties or fifties.

The person who works the land is often not the same person that owns the land. Over 50 percent of farmed land in Kansas is rented. The landowner may be a neighbor or relative, but many landlords simply own the land as an investment. Landlords are often reluctant to put up the money to invest in long-term improvements to the landscape. As long as basic soil conservation practices are being followed and returns are acceptable, they continue renting to the same tenant. Meanwhile, the tenant farmer has no incentive to invest in improvements since many leases are year to year.

Experience with the River Friendly Farm Program in Kansas has highlighted another important barrier to environmental planning—the need for a comprehensive financial picture. Ron and Vicki Jacques generously volunteered to be a host farm for one of the River Friendly Farm workshops in the fall of 2000. They raise sheep and grain crops in south-central Kansas and have two sons now in college who are planning to return to the farm. They have raised feeder lambs in the same 3-acre lot since the 1950s. Unfortunately, the previous owners located the lot right next to the county road, and the rain that drains off the lot goes into the county ditch and then into a stream. The oldest well on the farm is located in the sheep feedlot and is the animals' water source.

Ron hadn't tested the feedlot well or any others on the farm (including the house water supply) in years, but he was well aware that the present situation couldn't continue. He had dreams of building a new facility on top of a hill about a half-mile from the house. Since the old lambing barn was also in need of replacement, a new facility could solve that problem too. However, Ron hadn't had time to put pencil to paper to determine if the sheep operation could support moving the lots and building a new barn. There would be major expenses for water lines, power lines, as well as the building itself, along with feed bunks and a windbreak.

The Jacques received financial analysis services through Kansas State University's Department of Agricultural Economics. They discovered that moving the sheep would not necessarily be cost effective. However, county agricultural extension agents and NRCS technical specialists informed them that run-off from the sheep lot could be controlled by putting in a berm, waste-holding area, and biofilter instead of moving the lot. The River Friendly Farm notebook made them realize that they would need to test all the wells on their farm immediately and that their chemical and fertilizer storage facility would need to be upgraded to further protect the wells. Having both the financial and environmental assessments *together* made it possible for them to hold a family meeting when their sons came home for Christmas break. Now they had the critical information needed to discuss options for eventually expanding the farm to include their sons. On a real farm, the family, finances, and environmental protection are intertwined.

How to Proceed?

How do nonfarmers begin to collaborate with farmers or farm organizations to encourage conservation on private lands, as called for in chapter 1? A thorough critique of modern agriculture will not help. Many books written on sustainable agriculture begin with a list of all the problems, environmental and social, that result from modern agriculture. Then the solutions are listed, under the broad umbrella of sustainable agriculture. The critique is a natural beginning to an academic treatise, but this approach could be off-putting—to put it mildly—to someone who is farming now. It is probably one of the reasons the debate has been so polarized for the last twenty years. All of the ills of agriculture are not found on any given farm. Yes, one may have a leaky septic system or a stack of unrinsed pesticide containers in the corner of the machine shed, but is this *particular* farm really responsible for the dead zone in the Gulf of Mexico or for increasing antibiotic resistance worldwide?

Instead of tearing their practices down, whole farm planning builds people up by helping them become better informed about best management practices, their financial situation, and their options for the future. These qualities, in turn, can lead to environmental protection and restoration.

People need incentives to make significant changes in their way of life and business, and there are many ways of creating positive incentives for change. These could include direct payments for the farmer's time to do a whole farm plan, in-kind services such as financial consulting that would be helpful in the planning process, and cost-shares or grant funds for implementation of specific practices.

Creating a separate market for farm products raised in environmentally sensitive ways is known as "green" or "eco" labeling—though some farmers cringe at these words (see chapter 17 for a longer discussion of eco-labeling). Interestingly, farmers who have participated in the River Friendly Farm Program seem to have no problem with this phrase, and a group near an urban area in south-central Kansas is looking into developing value-added marketing, using a "River Friendly" label, in nearby urban areas.

Farm programs in place now include financial incentives to set aside land for grassland, woodlands, and wetlands and to create buffer strips along streams and other sensitive areas. The next round of U.S. farm policy may create even more incentives for farms that provide "multiple benefits" to society besides food and fiber. In the past, farmers were asked (via the federal payment reward system) to simply grow food as efficiently as possible. The "fence-row to fence-row" policies of the past are partly responsible for the high rates of soil erosion, fertilizer run-off, and habitat loss we currently observe. Due to the subsequent oversupply encouraged by these programs, we also paid these productive farmers not to plant certain crops in certain years—but without seeing any environmental benefits from these set-aside programs. A farm program that rewards farmers for providing benefits to society that we can all enjoy makes more sense than most previous payment schemes (see chapter 18).

For example, farmers that reseed crops to grassland or reduce their tillage operations via long crop rotations, for instance, could be paid for increasing the organic matter content in their soils, offsetting some of the carbon dioxide released in the burning of fossil fuels. Protecting air and water quality, wetlands, and wildlife habitat may also be part of the next farm bill proposed payment programs. Having a whole farm planning process in place will increase the likelihood that farm program benefits can be applied to each individual farm in a way that is helpful to the farmer and also provides the intended environmental benefits.

Composing a Farm: An Example of Whole Farm Planning

Keith Tollefson farms with his brother, fourth-generation farmers on the rolling hills south of Hiawatha, Kansas. Recently, his son and his brother's son have joined the operation. Their dairy barn, holding pens, silos, grain bins, and feeding area, are tucked into a valley, halfway between the crest of the hill and a creek that feeds directly into the Missouri River a few miles away. Their great-grandfather chose this site. His reasons included accessibility to water, protection from the north winter winds, and good air and water drainage. Now the family home is also located close by, and the whole operation sits only a few feet north of the paved road. Keith and his brother have been following Kansas State University recommendations for good crop production and have been careful stewards of the soil, implementing soil conservation practices on every acre they farm, including rented land. Terraces have been built and maintained, waterways planted to grass, and the stubble is left on the fields in the winter to catch the rain and snow. A diverse crop rotation allows them to spread the dairy manure on many acres, fertilizing and improving the soil.

What is the problem? Any manure that is not immediately scraped from the lot is subject to being washed off into a waterway that drains into the creek, about a half-mile away. The water from the milking parlor, built over twenty years ago, empties into this same waterway, as does the drainage from the upright silo. There is not enough land on the north side of the road next to the barn and lot to build a conventional lagoon or catchment area. Putting an additional pipe under the road and building a lagoon on the south side of the road are expensive propositions.

If there were ever a rainfall event that resulted in a fish-kill in the stream due to their dairy run-off, they would be subject to a $10,000 fine. However, moving the barn, lot, hay shed, and silo to a better site would cost between $100,000 and $200,000. With falling milk prices and dairy farmers going out of business, they are lucky—and clever—to be in business at all. There is not enough cash flow in this operation to support a move at this time.

The Tollefsons volunteered to be a training site for the Kansas River Friendly Farm program, for county agents, NRCS employees, and extension specialists. They completed a River Friendly Farm notebook, or checklist, and found that many of their farming practices met or exceeded best management practice guidelines. They also found that the location of the milk barn, feeding area, and manure- and water-handling practices did not. Through participating in the training, they received recommendations from a variety of specialists, including engineers, dairy, and soil management specialists and discovered that they did in fact have options. These options included a small holding area for milking parlor waste water, a biofilter area for lot run-off, and they also found that by diverting the majority of the water that flowed from the fields down the hill toward the barn, they would reduce the chance that the dairy waste would also be swept toward the small stream at the bottom of the hill. They also realized that eventually the entire operation needed to be moved to a new site, further up the hill, but that they could move in stages, spreading the major cost of new barns and construction over several years or even decades. They could start now, by planting trees for a windbreak at the top of the hill, north of the new proposed site. In addition, they were made aware of cost-share funds available from several state and federal agencies that would help them accomplish their goals, which include clean water, as well as providing an adequate income for their sons, and income for their retirement years. A quote from Bateson (1989) poignantly fits the Tollefsons' situation:

> Today, the materials and skills from which a life is composed are no longer clear. It is no longer possible to follow the paths of previous generations.

The future will not be easy, and each decision needs to be carefully considered and implemented, to balance environmental quality with their own quality of life and income needs. However, this is an example of how farm planning and assessment can help a farm family start down the road to solving what had previously looked like an untenable situation.

In the comfort of a distant office or conference ballroom, one can hold up the ideals of sustainability and demand that farmers change completely or face the cold cruel world of regulation. In practice, it is more productive to work toward sustainability the way one would coach an athlete for a marathon. Instead of asking him or her to risk bodily injury by running 26 miles on the first day of practice, the smart trainer starts with the athlete's own ability and level of endurance and works up from there. If we don't start working with farmers *where they are* now, we miss an opportunity to change the future.

To conclude with another quote from Bateson (1989):

> Once you begin to see these lives of multiple commitments and multiple beginnings as an emerging pattern rather than an aberration, it takes no more than a second look to discover the models for that reinvention on every side, to look for the followers of visions that are not fixed but that evolve from day to day. Each such model, like each individual work of art, is a comment about the world outside the frame. Just as change stimulates us to look for more abstract constancies, so the individual effort to compose a life, framed by birth and death and carefully pieced together from disparate elements, becomes a statement on the unity of living. These works of art, still incomplete, are parables in process, the living metaphors with which we describe the world.

References

Bateson, M. C. 1989. *Composing a Life*. Atlantic Monthly Press, New York.

Bidgood, M. 1995. *A Study of Actions by Simcoe County Environmental Farm Plan Participants*. Report to the Ontario Farm Environmental Coalition Office.

Janke, R. and D. Nagengast. 1999. "Environmental Farm Planning ." Kansas Sustainable Agriculture Series Paper Number 7, Contribution No. 99-390-D, from the Kansas Agricultural Experiment Station, Manhattan, Kans. Available online at http://www.oznet.ksu.edu/kcsaac/publications/sus_ag7.pdf.

McLean, M. 1999. "Commentary for the Soybean Growers in Chathan." Environmental Farm Plan Program, November 17. Available at http://www.soybean.on.ca/postit/commentary/archive90.html.

Norman, D., L. Bloomquist, R. Janke, S. Freyenberger, J. Jost, B. Schurle, and H. Kok. 2000. "The Meaning of Sustainable Agriculture: Reflections of Some Kansas Practitioners." *American Journal of Alternative Agriculture* 15:129–136.

Ontario Farm Environmental Coalition (OFEC). 1996. *Ontario Environmental Farm Plan*, 2nd ed. Ontario Farm Environmental Coalition, Toronto.

Chapter 15

After the Deluge

Integrated Watershed Management in the Red River Valley

Cheryl Miller

In 1997, one of the most devastating floods in American history occurred in the Red River Valley, an important agricultural region in the north-central United States. After a winter of heavy snows, an enormous snowpack blanketed the region, and when springtime came, it became a sheet of water stretching across thousands of square miles. For weeks, flood waters rose, froze in place during a sudden late-season blizzard, then rose again. Television images of people baling, sandbagging, and finally evacuating Grand Forks and other communities were seen around the world. Published estimates of flood damages topped $6 billion.

The flood was the largest of numerous floods in the Red River Valley during the 1990s, the wettest and warmest decade on record. Something akin to a state of siege had descended as massive storms rolled overhead, dumping rain onto a watershed already saturated and overflowing. Skirmishes moved up and down the Valley over proposals to construct or enlarge drainage ditches and dams. Nearly everyone got caught up in the cycle of disaster and repair: upstreamers fought downstreamers; city dwellers fought farmers; environmentalists fought local water authorities; government agencies fought internally and among themselves.

The battle over flooding—what was causing it and what should be done about it—reflected, in part, shifting attitudes about the dominion agriculture holds in the Valley. Conventional wisdom has it that society should "farm the best and conserve the rest." But because the Red River Valley supplies a substantial portion of the nation's sugar beets, spring wheat, and potatoes, few people—here or elsewhere—have expected or demanded anything else from it. To many people, the conversion of the region's extensive tallgrass prairie, rivers, wetlands, lakes, and woodlands to farmland could easily be justified by its agricultural bounty.

What was becoming clear, however, is that after a century of engineering the Valley's watercourses and lands for higher levels of agricultural production, the volatile weather patterns of recent years have pushed this "best" land into ecological, agricultural, and social ruin. Soil erosion is some of the worst in the nation, streams and ditches are choked with sediments, water quality is degraded, and what wildlife remains is restricted to tattered remnants of a bountiful wilderness. News reports from the late 1990s indicated that many Red River farmers were being pushed into bankruptcy by a combination of the sustained agricultural recession, loss of agricultural commodity supports, and a disease associated with wet soils.

In "The Land Ethic," Aldo Leopold (1949) wrote that avoiding social and environmental problems such as these required that land be managed as a "whole organism" of plants, animals, and ecological processes that had evolved over eons. "Man-made changes," he wrote, "are of a different order than evolutionary changes, and have effects more comprehensive than is intended or foreseen."

Leopold's philosophy is echoed today in holistic or ecosystem management, an emerging approach to land and water management that emphasizes the health and long-term sustainability of natural resources rather than output. This approach is based on a working understanding of ecosystem functions and their limitations and, frequently, some form of collaborative or consensus-based decision making about how ecosystems are managed. In the past several decades, ecosystem management has been applied to national forests and other public lands (Grumbine 1994), but is largely unheard of in farm country. Can such a paradigm be used in intensively managed regions like the Red River Valley? If so, what are realistic goals for managing water and natural resources in distressed watershed that are most entirely privately owned?

These questions—though never openly addressed—informed a remarkable discussion that began in the Valley in the aftermath of the 1997 flood when a formal mediation was begun to settle pending lawsuits over flood control. Participants in the mediation included representatives of federal, state, and local government agencies, watershed districts, farm interests, academia, and environmental organizations, including National Audubon Society whom I represented. The crisis in the Valley turned the focus of the mediation from the minutiae of regulatory law to larger economic, environmental, and institutional issues. This, in turn, led to a discussion about the need for a more comprehensive approach to managing water and, eventually, to an agreement on a set of goals and working relationships that align closely with the basic principles of ecosystem management.

This chapter presents a case study of how ecosystem management principles and processes were applied in the agricultural lands of the Red River Valley. It describes the formal flood control mediation and its final outcome, the Red River Basin Flood Damage Reduction Agreement. It also describes key elements of the Red River Agreement, signed in 1998, and relates lessons learned in the first years of implementation. This chapter concludes with an attempt to answer the questions posed above and reflects on the role of environmental activists in agricultural watersheds.

For the most part, this book describes how the practices of individual farmers can restore ecological services and wildlife habitat. This chapter focuses on how farmers and others within a watershed can work together on the landscape level to bring about improvements in ecological services that contribute to flood control. It also describes a process developed in the Red River Valley negotiations that is a step toward agroecological restoration and a model for resolution of other conservation conflicts in agricultural settings.

A "Vast and Beautiful Prairie"

Despite its name, the Red River Valley is not so much a river valley as it is the exposed bed of an enormous lake formed in the wake of a receding glacier ten thousand years ago. Glacial Lake Agassiz, named after Louis Agassiz, the great theorist of glaciation, was a 200,000-square-mile sheet of water, comprising parts of present-day Minnesota, North Dakota, Ontario, Manitoba, and Saskatchewan. Its lakebed is one of the largest level tracts of land in the world and, as the millennia drifted by, a vast prairie with a luxuriant growth of tall grasses 8 to 10 feet high grew up. Northern tallgrass prairie, part of a grassland stretching from Texas through Manitoba, dominated the basin and was interspersed with wet meadows, deep marshes, calcareous fens, and, in the northern reaches of the basin, extensive peat bogs. Oak and aspen savannas blanketed low beach ridges at the basin's edges and transitioned into dense pine and hardwood forests in glacial moraines beyond (Upham 1895, Ojakangas and Matsch 1982, Stoner et al. 1993).

The mix of habitat types, reshaped and renewed by frequent prairie fires, supported wildlife communities of legendary proportions. Bison and antelope, grizzlies, huge flocks of cranes and waterfowl, mink, otter, beaver, and hundreds of other species inhabited the open grasslands and forests and made this an ancestral hunting grounds of native peoples and in the 1700s a major outpost of the North American fur trade. From the journal of Alexander Henry, January 14, 1801: "At daybreak I was awakened by the bellowing of buffaloes. . . . I dressed and climbed my oak for a better view. I had seen almost incredible numbers of buffalo in the fall, but nothing in comparison to what I now beheld. The ground was covered at every point of the compass, as far as the eye could reach, and every animal was in motion" (USFWS 1997).

The extension of the railroads into the Red River Valley from Minneapolis, a major new milling center, led to the first great plowing up of the prairie in the 1870s. James J. Hill, president of the Great Northern Railroad, and others recognized the tremendous agricultural potential in the basin's level terrain, deep rich soils, and moderate summers. Hill was convinced that with government help, millions of acres of wetlands in the Valley could be drained and made attractive to the waves of immigrants moving west in search of cheap farmland. With his help, drainage laws were enacted and financing provided for "a thoroughly effectual and general system of drainage" (Prince 1997).

Within a remarkably short twenty-year period, the Valley's prairie and marsh environment was transformed into the world's premier spring wheat-growing districts. River segments were connected, enlarged, and straightened. Millions of acres of lakes and wetlands were drained. In 1911, a drainage project, touted as the largest ever undertaken, hooked a massive network of ditches through 600,000 acres of prairies, woodlands, and peat bogs into a small, meandering stream called Thief River. The failure of this network to turn peat into productive farm soils and the subsequent abandonment of much of it helped turn public opinion against the foolhardiness of many drainage entrepreneurs—for a while. Since that time, waves of private drainage have reoccurred in response to advances in drainage technology, or as market demand raised the value of marginal land, or during wet cycles (Bradof 1992).

Today as little as 1 percent of the prairie remains, most of it in tiny fragments surrounded by vast fields of sugar beets and wheat. Wetlands, except impenetrable bogs in the northern lake basin, have been reduced to 1 to 10 percent of presettlement acres. Natural river systems have been degraded by dam construction, channelization, loss of riparian habitat, and sedimentation. Woodlands—which comprised as much as 40 percent of the watershed and played critical roles in stabilizing soils, buffering streams, and moderating climate—are diminished by half (USACOE and MDNR 1996). The loss of these systems and the services they provide—water infiltration, water retention, water transport, and evapotranspiration—has resulted in cascading damages to farms, communities, and the environment.

The prodigious wildlife communities of the northern tallgrass prairie has long since disappeared. Free-ranging bison, the Great Plains wolf, swift fox, pronghorn antelope, and grizzly were extirpated early; other large carnivores, grazers, and predatory birds began to disappear by the end of the nineteenth century. For much of the twentieth century, other prairie species remained relatively common despite the tremendous habitat losses. In the past twenty-five years, however, many recently abundant species have entered deep and sustained declines (USFWS 1997).

Among the hardest hit are grassland birds, whose numbers are falling throughout North America. Species dependent on rare, declining, or vulnerable habitats—particularly the larger prairie/wetland complexes—for different stages of their life cycle are most at risk: greater prairie chicken, upland sandpiper, burrowing owl, bobolink, savannah, and Henslow sparrow. Other species of concern include migrating shorebirds, breeding terns, rails, waders, and the songbirds and raptors of the Valley's wooded habitats (Fitzgerald et al. 1998).

Flood Control: An Old Process Creates Conflict

The conflicts over flooding that arose in the 1990s revealed wide differences in how people view problems in the Valley and the recourse they seek. Depending on their point of view, people appeal to any of a number of laws and institutions that, too often, work at cross-purposes. A case in point is drainage and flood control.

Under Minnesota watershed and drainage law, flood control projects are initiated in response to petitions to watershed districts by landowners seeking relief from flooding problems. Generally, these and other affected landowners will pay most or all project expenses. Not surprisingly, they want to keep costs low. The district's engineers are, in turn, directed to design a cost-effective flood control project. The stress is on cost efficiency, at least until the project proposal is approved locally and sent to state and federal regulatory agencies for permits.

At this stage, a different set of laws comes into play: the National Environmental Policy Act, Clean Water Act, Endangered Species Act, Minnesota Public Waters Law, and Minnesota Wetlands Conservation Act, each administered by a different agency and each applying a different set of regulatory requirements. In addition, through the public notification process, a larger segment of the public hears about permit applications and has the opportunity to comment on them. Very often this larger public hears about potentially damaging proposals—say, to build a dam across a stream and impound water in a relatively natural area rather than dam a drainage ditch and flood a farm field—and starts asking questions. It frequently happens that years after a petition for flood protection has been filed and after tens of thousands of dollars have been spent on engineering, opposition is raised by downstream cities, agencies, environmentalists, or others who that demand the project be sent back to the drawing board.

This process, seemingly designed to create conflict, produced flurries of lawsuits. The U.S. government responded to public concerns about a series of flood control proposals by imposing a moratorium on new permits of flood control projects. Federal agencies ordered an environmental impact study (EIS) on potential cumulative impacts of thirty-three dams being planned by local watershed district boards in the Red River Valley. The study determined among other things that the proposed impoundments would inundate over 8,000 acres of wetlands, 1,650 acres of grasslands, and 630 acres of riparian forests. Although the EIS produced voluminous data on environmental resources and effects, it was not designed to address the underlying social and institutional issues and, when it was released in 1996, it was greeted by a new round of lawsuits. The study also provided a useful record of public opinion on the relative importance of the protecting or enhancing agriculture versus protecting these remaining natural areas:

> [A]gricultural production of this area has not reached its potential and is limited by lack of flood control. . . . Does activity on behalf of the human environment have more significance than a riparian blade of grass?
>
> —A local citizen

> The Red River Valley would suffer significant losses in wetlands and woodlots; many streams and rivers would also be adversely affected. The small benefit does not justify the wholesale destruction of the remaining riparian habitat found along these rivers.
>
> —A member of a local conservation organization

> The development of one of the world's most productive agricultural regions by its residents has been dependent on the ability to change the original conditions. The intended flood flow reduction projects are part of an ongoing initiative to preserve and enhance the agricultural industry, both production and processing.
>
> —An official with a regional water board

> The main benefit will accrue in the form of enhanced agricultural productivity. This is an important economic benefit that will be distributed mainly to landowners in the project areas. Projects for which benefits are distributed narrowly raise particular concerns about fair distribution of the project costs. Will those with the largest benefits incur a fair share of the costs?
>
> —An economics professor from a nearby university

> The Red River Basin can produce food and fiber more abundantly and with less environmental impacts than almost any other area currently under cultivation and certainly less than many areas that could be brought into production to replace Red River Basin farmland.
>
> —The director of a regional forum on the economy and environment

> Land use practices in the headwater areas need to be addressed. A regional water management plan must be formulated with the best interest of all the public.
>
> —A local conservationist

A New Cooperative Process: The Red River Agreement

In the months after the 1997 flood, the Minnesota legislature decided to finance a formal mediation to resolve the gridlock that had overtaken the flood control regulatory process. CDR Associates of Boulder, Colorado, which specializes in conflict resolution of natural resource disputes, was engaged to develop and facilitate this reconciliation process. After initial interviews to identify affected interests and potential mediation participants, CDR outlined a process that included a stakeholder roundtable and an interdisciplinary scientific and technical advisory team. CDR also established consensus decision making, and a timeline. The stakeholder panel included representatives of the Red River Water Management Board and local watershed districts, U.S. Army Corps of Engineers, U.S. Fish and Wildlife Service, Minnesota Department of Natural Resources, Minnesota Board of Water and Soil Resources, Minnesota Pollution Control Agency, Minnesota Center for Environmental Advocacy, and Rivers Council of Minnesota, as well as farmers, urban residents, and academics. I represented National Audubon Society and local Audubon chapters in the mediation.

The purpose statement adopted by the stakeholder group was "To reach consensus agreements on long-term solutions for reducing flood damage and for protection and enhancement of natural resources. Such agreements should balance important economic, environmental, and social considerations. Such agreements must provide for fair and effective procedures to resolve future conflicts related to flood damage reduction."

For eight months, during meetings and drives around the Valley, mediation participants developed greater understanding and appreciation of the people, the region, and the issues that needed to be sorted out. As often happens when consensus rather than "winner-loser" rules (and skills) are used, common ground was found and built upon. The common ground in this mediation was (1) the urgency of a major flood reduction effort; (2) an acknowledgment that solutions should not create problems elsewhere; and (3) realization that progress depended on turning opponents into cooperators and giving all a stake and a role in the program.

In December 1998, all participants in the mediation signed the "Red River Basin Flood Damage Reduction Agreement." The Agreement fundamentally changed how flood management will work in the region. It establishes a fifteen-year program, projected to cost upward of $250 million, to make significant progress on problems in the Valley.

Specific Goals for Flood Damage Reduction and Natural Resource Improvement

The magnitude of flooding in 1997 demanded not only that flood efforts go forward, but also that all affected groups—downstream cities, upstream farmers, environmentalists in distant cities—get behind them. To turn everyone into a stakeholder, two sets of goals were agreed upon.

Increased Flood Protection and Conservation

One goal was to increase flood protection for farms and communities. In its most important departure from the past, the Agreement calls for a significantly higher level of flood protection for farmland. Private drainage ditches typically are built to handle runoff from relatively small summer storms; they are not capable of handling floodwaters from bigger, less-frequent storms. Now, during an era of big storms back to back, the Agreement calls for a larger role for the public in funding protection for farmlands from ten-year storm events. The Agreement also calls for maximum feasible protection for human life, one-hundred-year flood protection for homes and communities, and increased protection for roads and other social, economic, and natural resources.

In exchange for actively supporting greater flood protection for farmlands, conservation interests were able to establish equally ambitious goals. All participants agreed to restore the natural characteristics and functions to the region's rivers and lakes; to restore 10 percent of the original wetland acres (as suggested in the North American Waterfowl Management Plan); to preserve and restore native prairies; to stabilize healthy populations of native species and provide connected, integrated habitat areas; to protect surface water supplies and groundwater recharge areas; and to provide increased recreational opportunities.

Integrated Watershed Projects

The Agreement calls for a comprehensive watershed approach to achieve these goals. Each of Minnesota's nine tributary watersheds within the Red River Basin will update its watershed plan to incorporate goals and principles of the Agreement. Not only will these plans further explicate how goals will be met, they will also include an annual process for evaluating and reporting progress toward the goals.

In promoting a comprehensive watershed approach, the Agreement states that flood control projects must take natural resources into account. Specifically, that means that watershed districts will design flood damage reduction projects to minimize environmental damage and, collectively, result in a net environmental gain. If, as expected, the projects are designed to address resource concerns and are sited to avoid impacts to remnant natural areas, such as river valleys, they have a much higher chance of passing muster in the regulatory process.

The Agreement's provisions put a premium on integrated flood reduction/natural resource projects such as river corridor restoration, setback levees, flood storage wetlands (wetland systems designed to allow 2 feet of floodwater bounce per ten-year storm), and a variety of conservation land uses or practices. The higher expenses of these types of integrated projects are expected to be covered by an array of state, federal, and private conservation programs.

Collaborative Process

The Agreement also calls for collaboration among government and private groups to develop and fund these integrated projects. Each of the nine tributary watershed districts forms a stakeholder team comprising the range of technical expertise, legal authority, and political powers needed to evaluate and fund multi-objective projects. The teams are guided by the Agreement and operate by consensus on all matters related to project development.

Finally, rather than attempt to amend any of the complex and politically sensitive policies governing water-related projects, the Agreement simply revises the project development process. Recognizing that each of the stakeholder groups holds part, but only part, of the power needed to advance or halt a project, the Agreement calls for early and direct stakeholder participation in project development. Instead of the old process that wasted time and money on projects that couldn't be permitted, the new process allows all involved parties to sit down together, examine the problem, develop alternatives, and reach consensus on a preferred option. As alternatives are examined, regulatory agencies identify information needs and flag issues that could slow or prevent permit approval. The all-important "alternative analysis" phase and the engineering and other technical work it requires are paid for through grants available to the watershed districts and their stakeholder teams. This fund greatly increases the chances that innovative approaches will be fully examined.

Lessons Learned in the First Years

The promise of this new approach is that all stakeholders stand to gain substantially more than each could have achieved alone. Landowners and farmers get higher levels of flood protection, with the public footing a significant portion of the bill. Conservationists gain strong local buy-in for restoration of natural areas and ecosystem functions, in large part because of the role such projects can play in reducing flooding.

During the first two years, the nine tributary stakeholder teams got organized, got acquainted, and selected their first joint projects. Some—especially those with professional staffs and a backlog of flooding problems to address—moved quickly. Others were more cautious and watchful of what the Agreement would bring. From observing them, it is possible to identify some critical elements needed for success in the early implementation stages of the Agreement: communication, a commitment to the learning curve, and funding.

Communication

The first thing stakeholder teams must tackle is the widespread distrust and lack of experience people have in working with those on the opposite side of environmental or flood control issues. After an initial year of experimentation, the original mediation participants reviewed progress and began work on additional guidance to teams. They learned that for the process to work, teams needed to

- Get representatives of all affected groups to the table and establish a high level of expectation and accountability in order to work constructively.
- Retain an outside professional facilitator experienced in consensus building and knowledgeable about the Agreement and issues it addresses. This is crucial to getting partisans to identify important concerns and to addressing them fully and objectively.
- Take time to build team spirit and skills, through tours, informal discussions, and healthy doses of humor. Build problem-solving and leadership skills of everyone in the group.
- Hold regular meetings and record decisions made and issues outstanding. Group consensus must be documented at each stage in the process (concept, engineering, construction, operations) to secure funding.
- Foster watershed-wide perspective through conferences, forums, publications, and the media. Keep the larger community informed and don't forget the old-timers in the coffee shops.

Scientific and Technical Issues

The learning curve on integrated watershed management is steep, long, and not yet fully built. A major investment in research and education in watershed management must be made. The participants agreed it was important to

- Retain a technical advisory group to provide expertise on hydrology, ecology, engineering, and other matters. A group of agency staff and private consultants should meet regularly to develop the technical foundation of the program through field research and reporting.

This group will develop objective measures to monitor and evaluate multi-objective projects.

• Involve academia. The Red River Institute is a newly forming cooperative program among colleges and universities in the United States and Canada. It will coordinate and fund applied research on watershed management.

• Spread information around; it is powerful. Develop and disseminate practical, pertinent information on water and natural resource management. The "Users' Guide to Natural Resource Efforts in the Red River Basin" is being published to provide easily accessible information on priority ecological concerns and objectives along with guidance on how to incorporate them into multi-objective projects (Miller 2001).

Funding

The program described above is designed to break the disaster/repair cycle that costs society billions of dollars in flood damages and to begin repairing natural systems and processes in the Valley. From this program, a number of principles became clear:

• All affected groups must help raise money needed and all affected groups have a voice in how it is spent. United project partners can access many sources of public and private money. Watershed district funds are very limited, are tax based, and cannot cover natural resource components of many projects.

• The process (team facilitation, education, and research) must be funded. It is as critical to a successful program as funding construction. The Minnesota legislature has appropriated $1 million annually to cover project team and other expenses.

• "Invest time and money early; save time and money later." In the past, watershed districts were reluctant to spend money on alternative analysis despite its central role in environmental permitting. This stage is now well financed and innovative alternatives can be thoroughly evaluated.

• Major watershed rehabilitation efforts benefit from programmatic, rather than project-by-project, funding. A federal Red River Initiative to fund cooperative flood reduction/resource improvement projects is being developed through the Army Corps of Engineers. If successful, it will provide a long-term stream of funding for projects fitting criteria established in the Agreement and should leverage matching money from state and other sources.

• New money should be sought. Long-term solutions to agricultural flooding will undoubtedly involve alternative economic uses of some flood-prone land. It is crucial that market opportunities and government policies give landowners viable alternatives for producing food, forage, biomass, or other products and services that are compatible with marginal farming conditions.

It's a Winding River

Americans, it is said, are adept at managing landscapes either as cornfields or national parks, and not very good at anything in between. More than in Europe or elsewhere, we Americans are inclined to intensively manage a piece of land or relinquish it as scenery rather than find some middle course (Bratton 1992). This dichotomy has led to an industrialized agriculture that uses up virtually all the life-support capacity of a region's soil and water and, in turn, dominates the region's political institutions and its economy. This degree of specialization increases vulnerability to environmental, economic, and other large-scale stresses such as the floods in the Red River Valley.

The Valley is, in fact, a textbook example of the tremendous vulnerability of specialized land uses and key institutions to environmental crises. In such situations, governing institutions either adapt to the demands of the time or are overwhelmed and marginalized by groups of people who step forward to tackle the crisis. Often, these informal collaborations by outsiders lead to fundamental overhauls of goals, policies, and organizational processes (Gunderson et al. 1995). The collaborations in the Red River Valley are leading not only to different kinds of flood control projects and a greater willingness to restore some wetlands and river corridors, they are also contributing to the restructuring of water management boards and processes and to a more shared vision of how this watershed can and should be managed. This vision does not include restoring the Valley to an eighteenth-century Eden, but no longer does it assume agriculture's domination in all matters related to the land and water. Though they're not using the "whole organism" view of Leopold, the collaborators are able to talk about the watershed as a system with essential and interacting parts and about the need for pragmatic goals to rehabilitate and protect them.

Environmentalists have an important role and stake in the changes occurring in the Red River Valley and elsewhere. The complicated problems and conflicts there will continue for years to come. But rather than standing on the outside and litigating our differences, conservationists can help invent the cultural changes that are ultimately needed for long-term environmental health. "Conservation is a policy problem," wrote Tim Clark in a 1994 paper on the biodiversity crisis, "and not merely a scientific concern, or a breakdown of stewardship values, or any other simple, reductionistic, cause-and-effect relationships. Such explanations can incorrectly tell conservationists that the problem is someone else's, and focus attention away from their own collective behavior, problem-solving style, and modus operandi."

Environmentalists need to learn how to "get on same side of table" as farmers and others struggling with a variety of environmental problems and cooperate in working through the social and institutional factors that drive environmental degradation. If we remain committed to key principles of identifying and involving all interests, basing decisions on objective information, and pushing for group consensus, real progress will be made in rehabilitating agricultural landscapes.

References

Bradof, K. L. 1992. "Ditching of Red Lake Peatlands during the Homestead Era." Pp. 263–284 in *The Patterned Peatlands of Minnesota*, edited by H. E. Wright, B. A. Coffin, and N. E. Aaseng. University of Minnesota Press, Minneapolis.

Bratton, S. P. 1992. "Alternative Models of Ecosystem Restoration." Pp. 170–189 in *Environmental Health: New Goals for Environmental Management*, edited by R. Constanza, B. G. Norton, and B. D. Haskell. Island Press, Washington, D.C.

Clark, T. W. 1994. "Creating and Using Knowledge for Species and Ecosystem Conservation: Science, Organizations, and Policy." Pp. 335–364 in *Environmental Policy and Biodiversity*, edited by R. E. Grumbine. Island Press, Washington, D.C.

Fitzgerald, J. A., D. Pashley, S. J. Lewis, and B. Pardo. 1998. *Partners in Flight: Bird Conservation Plan for the Northern Tallgrass Prairie (Physiographic Area 40)*. Version 1.0. August 4. U.S. Fish and Wildlife Service, Washington, D.C.

Grumbine, R. E. 1994. *Environmental Policy and Biodiversity*. Island Press, Washington, D.C.

Gunderson, L. H., C. S. Holling, and S. S. Light. 1995. "Barriers Broken and Bridges Built: A Synthesis." Pp. 489–532 in 1995. *Barriers and Bridges to the Renewal of Ecosystems and Institutions*, edited by L. H. Gunderson, C. S. Holling, and S. S. Light. Columbia University Press, New York.

Leopold, A. 1949. "The Land Ethic." In *A Sand County Almanac*. Oxford University Press, New York.

Miller, C. 2001. *Users' Guide to Natural Resource Efforts in the Red River Basin*. Red River Flood Damage Reduction Working Group, St. Paul.

Ojakangas, R. W., and C. L. Matsch. 1982. *Minnesota's Geology*. University of Minnesota Press, Minneapolis.

Prince, H. 1997. *Wetlands of the American Midwest: A Historical Geography of Changing Attitudes*. University of Chicago Press, Chicago.

Stoner, J. D., D. L. Lorenz, G. J. Wiche, and R. M. Goldstein. 1993. "Red River of the North Basin, Minnesota, North Dakota, and South Dakota." *Water Resources Bulletin* 29:575–615.

Upham, W. 1895. *The Glacial Lake Agassiz*. Monographs of the United States Geological Survey, Vol. 25. Government Printing Office, Washington, D.C.

U.S. Army Corps of Engineers and Minnesota Department of Natural Resources (USACOE and MDNR). 1996. Appendices G and H. *Final Environmental Impact Study of Flood Control Impoundments in Northwestern Minnesota.* Vol. 2. July. U.S. Army Corps of Engineers and Minnesota Department of Natural Resources, Fort Snelling, Minn.

U.S. Fish and Wildlife Service. 1997. Northern Tallgrass Prairie Habitat Preservation Area. Draft Environmental Impact Statement. USFWS, Washington, D.C.

Chapter 16

A Refined Taste in Natural Objects

Beth E. Waterhouse

It is daunting, after all the ideas and experiences covered by this book's previous chapters on agroecological restoration, to finally face and boldly address the underlying issue of motivation. Behind all the facts and experiential detail lies the question why. Why is it that some people choose to work in a closer partnership with the earth's forces and cycles, yet others cannot or will not do so? Why will one farmer protect a wetland while the adjoining neighbors are only able to see the world through rows of corn? I have closely considered these questions, calling the idea "environmental motivation," and for all my thinking and all the words I've read on the topic, it often comes down to one simple grand and mysterious force called love of the land.

Inside the age-old nature/nurture debate, one could say loving the earth is simply the way some of us are born. Is the short answer about environmental motivation then, that some people simply love the land and others do not? When I feel the growing disconnection between humans and their earth, I have to believe that there is more to be said. There must be more layers to the mystery of why.

The title of this chapter, "A Refined Taste in Natural Objects," is a phrase used by Aldo Leopold, who speaks of this mystery in his essay, "The Round River" (1966): "We need knowledge—public awareness—of the small cogs and wheels, but sometimes I think there is something we need even more. It is the thing that *Forest and Stream*, on its editorial masthead, once called 'a refined taste in natural objects.'"

Leopold later refers to this phrase saying, "I am trying to teach you that this alphabet of 'natural objects' (soils and rivers, birds and beasts) spells a story. . . . Once you learn how to read the land, I have no fear of what you will do to it or with it. And I know many pleasant things it will do to you." (Flader and Callicott 1991). Here Leopold gently calls for attachment. We know that Aldo Leopold believed this could be and was learned. He continually practiced close and informed observation with specialized training toward the recognition of patterns. This led him (and his children and students) to being more and more observant and apparently more and more attached to the miracles of interrelatedness inside a balanced ecosystem.

As this chapter plumbs the depths of environmental motivation and attachment to the land, it does so inside a deeper context. A disconnection between the eater and the farmer has become cultural. Too many citizens do not know or seem to care about what the farming countryside looks like or how farming practices may be polluting the groundwater. We've grown up accepting weed-free monocultures. The general public has grown up accepting farmland as a separate entity from "nature" in the purest sense. What might be the motivation for people to care for all of the land, even paying attention to the impact of food production on the land and to be as outraged about damaging farmland practices as any other improper land usage? What would it take to put farming back inside our understanding of environmental issues?

As we strive to reinstate agriculture inside the scope of "the environment," we must recognize that the two have never actually been separate. The false distinction has been a line of specialization. Leopold wrote in "Round River" that "the land is one organism" and he knew he meant farmland as well as nature preserves. Let farming as nature remain as the context for this book as well as for this chapter.

In this chapter, I will explore the foundation of loving attachment to the earth. With some life examples and quotes from other writers, I want to dig back behind someone's first commitment to a diverse landscape (wild or farmed) and continue asking why they chose to make that commitment, asking details of their personal motivation. If love is the final motivator, how do some of us come to love the land in this way? I will explore lifestyle elements that might lead a person to care to take a restorative approach to land use or farming—factors such as family or community heritage, freedom to wander, or time to observe. I will open a discussion of beauty and story and their unnamable powers to change the human spirit. In related opposition, I will ask us to think about the fears for our future as a very current part of humanity's burden. Plowing through those fears is part of what we must do to open a loving spirit about ecological issues.

It seems true that behind the simple answer of attachment lies a complex set of cultural steps or supports that might lead one to that point of personal motivation. If we rehearse these steps, we may be more inclined to duplicate them, to purposefully re-create and protect the human attachments that will defend this planet that we call home.

Living into Our Heritage

When individuals committed to working in the life sciences, on sustainable farms, or in ecology are asked about the turning point in each of their lives that led them to this commitment, they often speak of a parent or teacher, of a grandfather or grandmother. There was that first summer creating a garden, those years farming together, or decades of deer-hunting trips. The stories circle around to the individual whose enthusiasm for the land was contagious, often one who passed on land itself or the knowledge of where the wind blew and water flowed on such land.

One fine rainy June day, I traveled into southwest Minnesota to visit a farmer who is clearly modifying the social current and practicing farming on some of his land to restore prairie and increase biodiversity. I asked Tony Thompson of Bingham Lake, Minnesota, why he cared to do such things. Tony, who grew up on that acreage, could immediately name for me the moment his thinking had changed direction. He tells of the afternoon he was sitting in a roadside ditch, waiting to fill the corn planter. Tony, then out of high school but not yet focused on such things as ecological diversity, was sitting in a bed of blue-eyed grass. This tiny iris-like flower caught his attention and his admiration. By the time of our interview, Tony had also connected this epiphany experience to Aldo Leopold, quoting from Leopold's essay, "Marshland Elegy" (1966): "Our ability to perceive quality in nature begins, as in art, with the pretty. It expands through successive stages of the beautiful to values as yet uncaptured by language."

Yet there is more to this story. Supporting beauty as a capturing force on that day of spring planting was a family culture that spoke of such things and was willing to listen to Tony's thoughts. If a farm boy comes home one day excited over the observation of blue-eyed grass and hears words from an admired parent or adult to the effect that it is all foolishness, I believe he will stop immediately and turn toward another train of thought. Yet the opposite is also true. Young people can be turned toward wonder with shared enthusiasm and well-timed encouragement.

Tony Thompson showed me a small pocket notebook belonging to his grandfather, Reuben Johann Erickson. Reuben carried this little handbook from 1903 to 1909, between the ages of nine and fifteen. Its penciled script reveals a boy excited about the natural movement of birds and each year's return of his favorites. "Saw a long steady diagonal of geese go slowly flapping far overhead and that sight of it made the blood rush to my head," says Reuben in his journal. At age thirteen, this boy sketches the new bird, draws out notes of a new bird song, and even calls one song "a perfect revelry of enjoyment." Somehow Reuben Erickson's boyhood love of the land and birds (along with that of other observant and caring family members) is embedded in the family bloodline and stories of Tony Thompson. In that way, Reuben himself added to Tony's willingness to come home from corn planting that day and describe the blue-eyed grass.

Tony proceeded to learn about the prairie grasses—the associations and importance of grass as part of a plant community. Now, with a few key individuals, he runs Willow Lake Farm near Windom, Minnesota. His goals are both to be a successful farmer and to pass beautiful wildness on to the next generations, "always lessening our impact on the land yet remaining viable" as a farm.

Parents, connected to their land, beget children or grandchildren connected to the land if enthusiasms are shared, if experiences are joyful, and, I might add, if the land remains available to them. I know I'll refer here to my own connection to a beautiful spring-fed Minnesota lake called Big Thunder Lake. My grandfather's life on the shore of that lake came alive through my father's stories, and access to that homestead lakeshore maintains the connection.

Aldo Leopold at his renowned "shack" in Sand County, Wisconsin, became a poignant example of one man's desire to pass on a love and experience of nature to his students and his children. Said Leopold: "On this sand farm in Wisconsin, first worn out and then abandoned by our bigger and better society, we try to rebuild, with shovel and axe, what we are losing elsewhere" (Lorbecki 1996).

Leopold and his family spent decades of weekends at the Shack experimenting with ecological restoration. Nature writer Stephanie Mills describes the Leopold family's commitment to restoring their sand farm, quoting Leopold's eldest daughter, Nina Leopold Bradley:

> From being a dustbowl era ruin so barren that you could "see a mile in any direction" to becoming again a verdant tapestry of riparian woodlands, sloughs, oak and pine woods, prairies, savannas, and sedge marshes, the Leopold lands have enjoyed a second chance. The renaissance of this place . . . is a tale of applied hope and intelligence, and of the land's forgiveness. (Mills 1995)

Mills informs us that all five of the Leopold children would become natural scientists and four would be elected to the National Academy of Sciences. Perhaps this was due to the fact that, as she puts it, "the joy in the Leopold family life infused what they did on their land. They worked and played a lot, invented games, made music together. Nina Leopold Bradley talked about the restoration activity as being 'so rewarding that it becomes a part of you'"

(Mills 1995). Family love and acceptance were as important to this land/heart connection as scientific teaching.

The power of generational love, familial love for land mixed with love for each other on a particular landscape is obviously a deep motivator. John Berger, writing about French peasant life, eloquently articulates an ancestral connection: "I dig the holes, wait for the tender moon and plant out these saplings to give an example to my sons if they are interested and, if not, to show my father and his father that the knowledge they handed down has not yet been abandoned. Without that knowledge, I am nothing" (Berger 1979). Both Berger and Bradley talk of this oneness with land. "Without that knowledge I am nothing," says Berger's character. "It becomes part of you," says Bradley. This becoming one with the land is a mystery, but the words of Nina Bradley or John Berger are not chosen lightly. Land we love becomes part of us as we talk about it, spend our time learning about it, drink its water or eat its food, incorporate it into our dreams of our own future, and then teach others. Becoming one with the land leads a person to protect the land organism as if in self-protection.

Living inside our heritage on the land, we are also held accountable by family and duty and memory. We are given a legacy to uphold; a grandfather's boyhood joy in the return of the warbler resounds in our own ears. We become connected to the land, one with the land, in ways we can hardly put into words. Yet this oneness leads us, decision by small decision, toward a set of behaviors that may later be called restoration.

I must add that I hope and expect that elements of this "family heritage" can be replicated for those who are personally disconnected from land ownership. Good teachers are key here, field trips and experiences on and with the land help replicate this—love and "oneness" can, perhaps, be taught as students have fun with the land and come to know its biological patterns and the magic of the science behind its patterns.

Connecting Our Spirit to the Land through an Appreciation of Beauty and Story

Beautiful blue-eyed grass, for Tony Thompson of Willow Lake Farm was the first motivating impulse of "pretty" that he remembers as a turning point. The tiny blue flower startled him; it got his attention. He brought it home and his family heritage helped make it safe and inviting for him to study it, but its beauty was the starting point. When I've mentioned my writing theme to farmers, they remind me to remember beauty and to talk about the aesthetics of diversity. The vireo sings; the heron lifts off the shallow pond in that corner of the field; the prairie grasses turn a dozen colors in autumn.

Not being satisfied to simply name beauty's influences, I continued to ask why it might be that beauty would affect our lives and strengthen our connection to the landscape around us. I have long been motivated by natural beauty. It restores my soul on the day's walk to the lake. But how is it that beauty works down inside us to bring things to a state of harmony? I believe it is physical, like the power of a long and full breath.

Scott Russell Sanders tells a loving story about the beauty of his daughter's wedding day. He happens to be comparing the wedding photos with photographs of nebulae taken by the Hubble Space Telescope, and he lifts it all into delicious abstraction, calling the effect of beauty a "hum of delight."

> Pardon my cosmic metaphor, but I can't help thinking of the physicists' claim that, if we trace the universe back to its origins in the Big Bang, we find the multiplicity of things fusing into greater and greater simplicity, until at the moment of creation itself there is only pure undifferentiated energy. . . . I think the physicists are right. I believe the energy they speak of is holy, by which I mean it is the closest we can come without instruments to measuring the strength of God. I also believe this primal energy continues to feed us, directly through the goods of Creation, and indirectly through the experience of beauty. As far back as I can remember, things seen or heard or smelled, things tasted or touched, have provoked in me an answering vibration. . . . I sense in these momentary encounters a harmony between myself and whatever I behold. The word that seems to fit most exactly this feeling of resonance, this sympathetic vibration between inside and outside, is beauty. (Sanders 1998)

Sanders proceeds, as he puts it, to "give biology its due," and he explores beauty even more:

> It may be that in pursuing beauty we are merely obeying our genes. It may be that the features we find beautiful in men or women, in landscape or weather . . . are ones that improved the chances of survival for our ancestors. Put the other way around, it's entirely plausible that the early humans who did not tingle at the sight of a deer, the smell of a thunderstorm, the sound of running water, or the stroke of a hand on a shapely haunch, all died out, carrying with them their oblivious genes.

I'm convinced it's more a matter of spirit than genetics; however, Sanders' last argument is intriguing. What fun to imagine, through the ages, the effects of beauty on generation after generation. Beauty was likely the topic of conversation around a cave's open fire, or inside a sod hut on the prairie, as it still is around the oak kitchen table of a family in farm country.

Each person defines beauty in his or her own way. We all know it when we see or hear it. But what of beauty misunderstood? The creative farmer knows the "aesthetics of diversity" mentioned earlier, but the lay person may not. Are we prone to protect what in our culture seems most obviously "beautiful" and stop there? Acres of waving corn used to be beautiful to me until I came to know the damages caused by a lack of diversity. If this is learned rather than immediately perceived, how difficult it then is to include diverse farming inside the scope of nature's beauty. Naïve or not, I come to believe it can be achieved, that eaters everywhere can come to understand diverse beauty on the land and come to connect the way we eat with a beautiful landscape.

My own reaction to beauty is visceral. I find that a small sound involuntarily comes into the base of my throat. When we find ourselves in beautiful places and open our eyes and ears and hearts to a sensory beauty around us, we reconnect breath and spirit to the land and its living web.

Story, like beauty, is a powerful connector. A good story is honest, universal, and sensory. Through a story we feel terrified, wet, cold, lonely, hopeful, rejuvenated. We step out of our own skins and assume the hide of the lion or the jaguar. We remember the fear of being lost, the exhilaration of finding home, of knowing our own mother, of feeling courageous. Stories refuse to reduce the sensory to meaning. They take us on a journey through sympathy or justice and leave us wandering our way back home through our own briars and brambles. As we read (really read) a story, we travel in our imaginations and take a break from the intellectual. We disconnect from the act of "solving" (with our minds) and connect to the act of "imagining" (with our senses.)

George Wuerthner, a wilderness explorer and writer, clearly expresses this power of story, saying, "Science may tell us that cows trampling a riparian zone results in fewer fish in our streams, or that logging old growth forests causes spotted owls to go extinct, but whether that is perceived as a problem or not depends upon one's values—and those values are shaped by the stories we use to guide our lives" (Wuerthner 1998).

My father's stories about living close to the land have brought that history and that land to life for me. In story, the details are shaped by the storyteller's values and values transferred through sensory detail. So we all continuously go about learning and shaping, shaping the learning—both sensing and knowing. In this process, we are building our values and "refining our taste in natural objects." Through story, in that internal narrative, we are learning our way home.

Rethinking Our Relationship with Time

Time and beauty go hand in hand. "Stop and smell the roses" is a common reminder, and we need to slow down for beauty. I sit still on the end of the dock for the whole sunset. Sure beauty can come in flashes, yet for beauty to embed itself, to have its full impact on our souls, there must also be slowness or repetition over time.

I know that it was a ten-day horseback ride with my dad that first got me to rethinking my relationship with time. One summer we committed to a ride from his home to his birthplace (near that same Thunder Lake mentioned earlier) diagonally through the state of Minnesota. Mostly, the horses walked for 270 miles. The oak tree on the plain would appear, small on the horizon, a growing image that would accompany us for hours. The birds or butterflies moved at our pace and we experienced a natural animal rhythm with life around us. The return trip, in a car pulling horses in a trailer, was actually frightening at 60 miles per hour, trees and birds flashing past.

After this trip, I began to see speed differently, and it seemed only to increase choices, to add quantity to our lives, but not necessarily quality. Efficiency, if it cut the artistic or sensory nature of my task, became a cost, not a benefit. Fifteen years later, I reorganized my life with the full understanding that a dollar saved is time in my pocket. Being has become as important as doing, for extended periods of time.

The perception of time shifts around for us dramatically by our focus on it or by its context. It drags waiting for a loved one's return or a diagnosis; it flashes by on a sunny afternoon's canoe trip. That canoe trip later lengthens again as we share stories about it, look at pictures of it, and retell it. We can manipulate our relationship with time.

Jacob Needleman, professor of philosophy and comparative religions at San Francisco State University, lends detail to this discussion of time.

> We're coming to the end of a hundred years or more of devices that were invented in order to save time. What has become of time? Nobody has enough time anymore. . . . We are a time-impoverished society. We have lots of material things, but we have no time left. Human time has disappeared, and we're in animal time. Or vegetable time, if you like. Or mineral time. The time of computers. The time of things. Of mechanical devices. . . . It's the New Poverty. (Needleman 1999)

Needleman's analysis reminds me of what I call the empty house syndrome—gorgeous huge houses that cost a mint to build and maintain sit empty all day because people have to work long hours to keep them. I who may never own a house again get to walk slowly by these empty houses simply because I need less space and have chosen to accept the gift of time. The house that might be so glorious to spend time inside demands absence, and the irony of this goes unnamed by most of the American working public. I believe that the lack of time named by Professor Needleman is directly related to the myth that runaway soul-less capitalism is the inevitable way of the American democratic future.

All of this relates to the earth in dramatic ways. Given time we can sit amongst the blue-eyed grasses suddenly noticing their beauty. Given time, (and trees) we can observe the patterns of nature, the return of the migrating birds, the growth patterns inside our garden or our neighboring woods or prairie. It takes time to come to grips with the changing of these patterns. It takes time to figure out how our own lifestyles can change so they impact the earth less.

A story told about the Australian aborigine goes something like this: A walkabout trailed late into the afternoon when the long single-file line through the Outback suddenly stopped. Questions traveled up to the front of the line, "What's wrong? Is there something blocking our trail? Is someone hurt?" But the answer came back that nothing was wrong; no one was hurt. "We have traveled far and fast today," was the reply, "and it is time to let our souls catch up."

We gain perspective from a slow-time gift to ourselves, and we lose ecological perspective with speed. In all our losses due to this high-speed culture of ours, I believe our relationship to Nature is one of the first and greatest losses. We need walkable streets, time to walk them, time to spare, time even to "waste" as our spirits sit amidst prairie flowers—as our souls catch up.

Wandering in Safety during Childhood

Wandering may be one of the simplest ways in which land-based experiences reach into our souls. There is a freedom and an emotional learning in wandering, in being lost and finding our way, which teaches us like nothing else. I've come to believe that one powerful motivating factor leading to a later commitment to environmental restoration may be a child's own freedom to wander. At a 1998 conference, with a group of individuals dedicated through school or career or habit to ecology, I asked about wandering. Had they wandered as children? A majority of them said yes.

I did too. My cousin and I would make a couple of sandwiches and take off via rowboat to the Lake's lagoon or by foot into the woods. Once we got lost; occasionally we were afraid. We always got home, and I know we created deeply rooted body memories along the way.

The deep-belly awareness that we are in charge of our own lost or foundness, the fear now and again of having made a wrong turn, and of being personally in charge causes us to pay close attention. It can connect us emotionally or through our senses to the land, for life.

Wandering is also directly connected to living our heritage. I am thankful that our moms would let my cousin and I go for a day at a time, only asking us to return for dinner. They knew we would find our way, and their trust that we would be safe instilled a matching trust in us. Wandering opened up before us a zillion creative possibilities about ways to spend a day, about the plant life or animal life in front of us. We learned to trust nature and to trust our choices.

It is not only the emotion in wandering or the trust built through wandering, but the memory of it that stays with a wanderer. We remember our love of the white birch and burr oaks or recall the pungent smell of freshly cut hay. Those free and wandering days create retrievable memories so that later the sights or smells in a field experience carry with them that old feeling of freedom.

Plowing through Our Fears

I truly don't know, some days, what it will take. I'd like to believe that we can recognize our own incremental losses or learn from huge ecological mistakes in other places (like the 1980s death of the Rhine River) and that fresh disaster is not our only motivator. The following stark statement was buried in a small column in the back of my daily paper recently: "The current rate of decline in the long-term productive capacity of ecosystems could have devastating implications for human development and the welfare of all species." You might think such a statement would get front page coverage. Not only does the media seldom do justice to such dire facts, but the culture seldom will hear it. We are living along here, imagining that we can undo our mark on the planet, imagining that hydro-cell cars or some new genetic decoding will reconfigure all the negative predictions. Yet many carry the burden of knowing better.

Aldo Leopold described this burden: "One of the penalties of an ecological education is that one lives alone in a world of wounds. Much of the damage inflicted on the land is quite invisible to laymen . . . [and we exist] in a community that believes itself well and does not want to be told otherwise" (Leopold 1966). I know many, especially those educated and with experience observing nature, who are markedly afraid of what we as humans have done to this planet in upsetting millennial cycles of earth, water, air, or sunlight. And blame is only a short-term salve.

Our relationship to this planet is, indeed, way too big a question anymore. People do not know where to start and some no longer trust that their small recycling or buying habits make a whit of difference when the very weather patterns are changing. It takes a great deal of openhearted courage, at this the turn of our century, to act in seemingly miniscule ways against vast corporate or development forces that would industrialize nature or pave the planet. The challenge of scientist, ecologist, or mystic is to talk, to refuse isolation (refusing to "live alone in a world of wounds") and somehow to act upon our knowing with even the smallest steps.

It is time we act over fear and pull our knowing up into some light of discourse. But how? Joanna Macy, scholar of Buddhism and deep ecology, describes how her friend, forest activist John Seed, deals with despair. "When I feel despair," he said, "I try to remember that it's not me, John Seed, who's protecting the rainforest. The rainforest is protecting itself, through me and my mates, through this small part of it that's recently emerged into human thinking." Macy also proceeded to discover a kind of group process that could move us beyond our shrunken human self-interest. She and John came up with the Council of All Beings (Macy 2000), a ritual now practiced the world over, a simply structured rite in circular form, where people create masks and speak on behalf of other life forms.

> "I am Wild Goose," says one. "I speak for migratory birds."
> "I am Wheat, and I speak for cultivated grains."
> "I speak for Weeds, a name humans give to plants they do not use."

Macy conjectures that humans imagine themselves to be in a place of fear and isolation. "When we get out of own identities," Joanna Macy has learned, "we allow ourselves to feel."

Beyond imagination, even, may be the sense of the sacred. Annie Dillard (1982) encourages us:

> In the deeps are the violence and terror of which psychology has warned us. But if you ride these monsters deeper down . . . you find what our sciences cannot locate or name, the substrate, the ocean or matrix or ether which buoys the rest, which gives goodness its power for good and evil its power for evil, the unified field: our complex and inexplicable caring for each other and for our lives together here.

Collaboration, ritual, science, yes, but even deeper that "ether which buoys the rest." Dillard takes us straight to our caring natures. Putting on the mask of the badger or otter is a shortcut to spirit. I hear myself as Otter, saying, "Give me unlittered shoreline; give me space to play. Come play with me and understand that fragile place where water and land meet." When my Otter's mask is on I want to cry out, to speak truths that my middle-class Minnesotan self might know but not boldly declare.

It's no coincidence that art is a part of embracing these issues. Fear is not easy to face head-on. Art—poetry, nature writing, sculpture, music—takes us out of our heads and into our bodies. Art addresses things around the corner of our intellect, opening us through rhythm or sharp poetic honesty, through masks, or color, or form. Art lets us say what articles stifle. Some environmental groups are now recognizing the need for writers or artists in their midst.

Using art, crossing disciplines, working together, all these things help us avoid the Mobius strip of fear that comes embedded in environmental conversations. Art provides another language, collaborative energy provides unity and hope. We look to our love of the land and see where it has come from, and open those options to our children. We might then begin to face our fears and open to our most creative selves as we redesign the future.

This chapter has cracked open the topic of motivation on several levels. In various ways, the individual is motivated to initially ask the question, "Why restore?" The greater society—whole groups of people, educational institutions and agencies—must also be motivated to see the landscape as a whole, to refuse to accept agricultural landscapes as sacrificial zones (a problem outlined in the first section of this book), but instead to pull farmland back into environmental considerations. We might bring farmland into our culture of beauty, into our hearts as part of nature—as land to be cared about, land to be loved.

I've come full circle in this discussion back to the strength of love. Can we refuse to diminish this powerful word as "too soft?" Can we nurture elements in our lives or our children's lives that become the foundation for a loving attachment to the earth? With beauty and story, with intention and creative leadership, I believe that we can.

References

Berger, J. 1979. *Pig Earth*. Random House, New York.

Dillard, A. 1982. *Teaching a Stone to Talk*. Harper & Row, New York.

Flader, S. L., and J. B. Callicott. 1991. *The River of the Mother of God and Other Essays by Aldo Leopold*. University of Wisconsin Press, Madison.

Leopold, A. 1966. *A Sand County Almanac with Essays on Conservation from Round River*. Oxford University Press, Oxford.

Lorbecki, M. 1996. *A Fierce Green Fire*. Oxford University Press, Oxford.

Macy, J. 2000. "Wild Goose." *Orion: People and Nature* 19:12–14.

Mills, S. 1995. *In Service of the Wild*. Beacon Press, Boston.

Needleman, J. 1999. *Time and the Soul*. Doubleday, New York.

Sanders, S. R. 1998. *Hunting for Hope*. Beacon Press, Boston.

Wuerthner, G. 1998. "The Myths We Live By." *Wild Earth* 8:6–7.

Chapter 17

Food and Biodiversity

Dana L. Jackson

"The Splendid Table" is a popular radio program distributed by Public Radio International, featuring Lynne Rosetto Kasper. She's the author of several cookbooks and an expert on Italian cuisine. Callers ask her questions about cooking techniques and where to find specialty cooking equipment. Kasper favors fresh, locally grown organic produce and talks excitedly about farmers' markets, but she also informs listeners about exotic ingredients in dishes from all parts of the world and the gourmet stores where they can be found. Those who call in seem to be interested primarily in an elegant cuisine, in creating a gourmet table. Callers generally aren't working parents seeking ideas for evening meals they can fix for their families quickly and economically.

Unfortunately, many of those working parents are feeding their families from the "fast food table," either in restaurants or copycat meals at home. After a school lunch provided by a fast-food hamburger chain, children often eat an evening meal delivered by a pizza place or are served microwaved french fries and chicken nuggets that parents picked up off the supermarket's frozen food shelves. Prepared salad mixes and peeled baby carrots from the supermarket may add some nutritional diversity to their menus, but for the most part meals are just tasty calories.

Another, probably smaller, group of consumers strive for the "safe food table." You can see these serious shoppers at supermarkets on the weekends, reading labels diligently to weed out foods with extra additives, food coloring, and cholesterol. They also shop the food co-ops for organic dairy products or milk from cows not treated with bovine growth hormone and pay incredibly high prices for out-of-season organic vegetables and fruits. Concerned with convenience as well as food safety, these shoppers are choosing boxed, organically grown and processed vegetarian meals. And they are buying fresh meat that is either labeled organic or "natural." Some of them have become consumer activists, who, for example, urged the United States Department of Agriculture to make the proposed rules for organic certification more stringent. They don't want to buy foods that have been genetically engineered, and they favor clear warning labels on products containing genetically engineered ingredients. Such consumers are primarily concerned about how food products affect their bodies, and only secondarily, or not at all, about how producing or processing food affects the whole environment.

Most consumers don't fall into any of these three categories but into a larger category of people looking for low prices and good value from a mainstream supermarket near them, people dependent upon the conventional food system. Their task is to provide for the "family table," which means affordable, easily prepared, and satisfying foods that family members will eat. Working parents will need ingredients for evening meals—not all of them serve fast food—appetizing cereals for kids who really don't want to eat breakfast, and sandwich makings for the sack lunches they take to work, plus snacks for the whole family. Almost all of the items tossed into their grocery carts, in spite of their labels, are products of a few major food corporations (Grey 2000). Fruits and vegetables are generally shipped in from California and Florida, Central or South America. Meats can't be traced to their origin but have come through a pipeline of large producers, buyers, processors, and distributors. Most supermarket shoppers vaguely know that crops and livestock are produced in the countryside somewhere and go through a number of handlers before being delivered to groceries and restaurants. However, few still have grandparents or uncles on family farms and can remember visits when they fished in the creek with cousins or rode on the combine during harvest. They have no ties to farmland and just want the food system to keep their favorite products easily available at low prices.

Whether they know it or not, the choices consumers make when they fill their grocery carts profoundly affect family farmers, rural communities, and rural landscapes. Although it seems that shoppers in supermarkets have a wide variety to choose from, in reality a small number of companies own the brands and control our food. What is available to us is more the result of business decisions about production, processing, shipping efficiency, and shelf life than it is about nutrition, the local economy, or stewardship of the land. Whether people shop for the gourmet table, the fast-food table, or the family table, they are generally buying food produced in an industrial system and shipped to their groceries from another state or country.

This book is about restoring a relationship between farming and the natural world, and this chapter is about restoring a relationship among eaters, farmers, and the natural world. The changes in our food system that are needed to return a greater level of biological diversity in rural landscapes require an impetus from the market, from consumers shopping for the "sustainable table." Consumer demand for food from sustainable farms must be built, and the best place to start is with people who are concerned about the environment but haven't made the connection between their grocery list and the endangered species list yet. First they must be made aware of the environmental and social costs of our conventional food system and then informed about alternatives to standard supermarket products and how they can buy them.

The Costs of Our Food System

People like the bargains they find in supermarkets, but supermarket prices don't really reflect the true costs of supplying food for the family table. They aren't aware of how much the food system is propped up with their tax dollars. Through farm programs passed by Congress and administered by the U.S. Department of Agriculture, farmers receive direct payments for growing certain commodities, such as corn and wheat. Farmers produce high yields using the tools of industrial agriculture—genetically engineered seeds, chemical fertilizer, herbicides, insecticides and large, sophisticated field equipment—which cause surpluses that drive prices below the cost of production. Then the taxpayer makes up the deficit by paying subsidies to farmers. The *Minneapolis Star Tribune* reported on October 3, 2000 that twenty-eight billion dollars from U.S. taxpayers were paid to the producers of major commodities in 2000, accounting for more than half of all farm income. These subsidies then enabled producers to pay the agribusinesses supplying the chemicals and machinery needed for high yields. Agribusinesses always lobby Congress for farmer subsidies because it's a pretty good deal for them to have taxpayers guarantee their profits each year.

Taxpayers help out others in the food system besides farmers. Since many of the surplus commodities must be exported, the taxpayer also pays for "export enhancement" programs to help large grain companies and food processors market their products abroad. And, of course, shippers benefit from the taxpayer-supported transportation infrastructure of roads and highways and locks and dams on rivers where barges move grain. Taxpayers support the land grant colleges of agriculture and the state and federal agencies that produce research to continually increase yields of commodities and process them into packaged food. They support national and state environmental agencies that are supposed to regulate farming practices to protect the environment, and public health agencies and the Food and Drug Administration charged with assuring the safety of food.

The environmental costs of our industrial food system range from actual dollar costs for pollution control and mitigation to unquantified costs in loss of natural amenities. The loss of clean fishing streams, trees, wild flowers, and songbirds in the countryside not only decreases the quality of life for residents of rural communities, but also limits their potential for future economic development, because such areas are less likely to attract new residents or businesses. The loss of biological diversity and ecosystem processes will also reduce the long-term ability of the land to be agriculturally productive (CAST 1999).

The average person doesn't know about these losses, or that their tax dollars support farming systems that cause them, but neither does the average environmentalist. People who worry about species extinction in the Amazon due to deforestation and the loss of biological diversity in the ocean due to global warming may be unaware of how food production closer to home disrupts the natural world. They may know about the decline of songbird habitat in Latin America but not know about the decline of habitat for those same birds when they migrate back to the Midwest. They are likely uninformed about the connection between Upper Midwest farming practices and the zone of hypoxia in the Gulf of Mexico where excess nutrients cause eutrophication and an oxygen-poor environment that cannot support most aquatic organisms. When it comes to buying food, environmentalists resemble the rest of the population: some shop for the gourmet table or the safe table, some buy fast food, but most shop for the family table, and they are often just as unaware of the connections among food, farms, and nature as their neighbors in the checkout line.

Can the Organic Table Be Sustainable?

Whichever kind of table consumers shop for, they generally have one priority in common: personal satisfaction. Food must satisfy their appetite and hunger at a price they can afford. It must taste good and look good, be safe and easily available. However, consumers who want to lessen their contribution to the environmental costs of food production must have additional priorities. They will look for foods that were grown and processed in ways that protect soil, water, and wildlife. Shoppers for the sustainable table will also consider social factors, such as worker safety and impacts on local community economics and quality of life. For decades, environmentally conscious consumers have chosen to buy organic food because they believe it is better for them, for small family farmers, the environment, and society at large. But, ironically, the consumer who buys USDA certified organic food these days may not be supporting these assumed benefits of an organic food system.

Rachel Carson's *Silent Spring*, published in 1962, directed people's attention to the down side of using pesticides in modern agriculture. By the first Earth Day in 1970, most of us becoming active in the environmental movement were very aware that chemical pesticides killed birds and that DDT was responsible for the soft-shelled eggs of eagles that threatened their extinction. People naturally began to think about how poisonous residues on food could affect their personal well-being. The impact of agricultural chemicals on the environment was part of the litany of issues recited on that first Earth Day. Activists urged consumers to buy organic food as a protest against the ubiquitous use of pesticides in farming.

Interest in organic gardening and organic foods burgeoned during the 1970s as part of the "back to the land" movement, and alongside Rodale's *Organic Gardening Encyclopedia* on our shelves, cookbooks appeared that stressed cooking with whole foods and natural ingredients, such as honey instead of refined sugar. My friends and I started making bread and granola with organic flours and grains at home. To increase the availability of whole foods and organic foods, we organized the Prairieland Food Co-op in Salina, Kansas, where we met monthly to order items such as beans, grains, flours, nuts, and raisins in bulk. This was happening in many towns and cities all over the United States, and soon an infrastructure of regional warehouses to service coops developed. Storefront cooperatives and health food stores selling organic food became permanent fixtures in business districts during the 1980s and 1990s.

The River Market Cooperative in Stillwater, Minnesota, where I shop today, is more like a glistening supermarket, than the dark old storefront where I bagged my food in the 1980s. The organic market has grown over the past decade at about 20 percent per year (DeLind 2000), offering diverse kinds of organic processed foods in boxes that compete for sales with whole foods in bulk bins. Customers are more likely to be sharply dressed professionals, not counterculture symbols with tattered jeans and ponytails. They exercise regularly at their health clubs and tend to seek organic foods more out of a concern for personal health than an environmental consciousness. Their willingness to pay more for organic food, however, can reward local organic farmers (the River Market features produce from local growers) with a price bonus for building organic matter in the soil and controlling insect pests without poisons.

During the 1970s and 1980s, organic vegetables and fruits could only be found seasonally at local farm stands, farmers' markets, and some health food stores. As demand increased in the 1990s, more began to be shipped from California year round and appeared on produce racks in Midwest and eastern states. Because consumers didn't know the growers, they needed a label certifying that the food was produced and processed according to organic standards. Organic farmers needed the label to assure they would receive the special premium accorded organic foods. Voluntary grassroots certification programs arose in thirty-one states, developing standards, training inspectors, and granting certification (DeLind 2000). Then, in 1990 the National Organic Foods Production Act was passed for the purpose of establishing uniformity in standards that would make it easier to ship products across state lines and national borders. After ten years of work by the National Organic Standards Board and an extensive public comment period followed by negotiations, the U.S. Department of Agriculture finally released a set of standards in December 2000 that will allow qualifying producers to use a label that says "USDA Certified Organic."

Consumers have generally assumed that small, family-sized farms grow organic food, but that is changing with its growing popularity in supermarkets and the new national standards. The cost of organic certification may be an obstacle that prevents small farmers from applying for the USDA Certified Organic label, while it will be an insignificant business expense for large companies. Also, small organic vegetable producers may lose their share of local markets, just as small conventional producers did, because they can't compete for price, quantity, and consistent supply against corporate producers. For example, fresh salad mixes of baby greens were a novelty originally marketed by small organic farmers in local markets; now Dole Corporation sells four kinds of organic salad mixes in grocery stores throughout the nation. And just outside Fresno, California, there is a 2,000-acre organic vegetable farm called Greenways Organic, which is just part of a 24,000-acre conventional operation (Pollan 2001). The scale advantages of large farms, vertically integrated with processors and shippers, make it hard for small farmers to compete, even if they are certified organic producers.

The willingness of customers to pay more for organic products has not escaped the notice of the food-processing industry. General Mills Corporation purchased two successful organic brands in 2000, Cascadian Farms (fruit jams) and Muir Glen (tomato products), and more corporations are expected to enter the market soon with their certified organic products.

Food Labels for a Sustainable Table

Consumers' purchases for the sustainable table should support—as much as possible—local family farmers and rural communities, worker health and safety, humane treatment of farm animals, and protection of the natural environment for the benefit of the larger ecosystem as well as for the human community. These have been ethical dimensions in the local and regional understanding of organic agriculture, and some forms of them are found in private or state certification processes but not in the national organic standards. The new national standards for organic certification describe the basic growing and processing methods acceptable for products using the organic label but do not address the larger social and environmental considerations for a sustainable food system (DeLind 2000). To provide food for the sustainable table, we need more than a list of dos and don'ts that we expect producers and processors to follow.

Farmers in different parts of the world have developed other kinds of labels and certification processes besides organic. They reflect the broader social and environmental values we associated with organic before the national standards were approved. Specific local or regional labels and eco-labels have sprung up for all kinds of products in the United States and Europe (Chasteen 1999, IATP 2000, Nadeau 1998) with an appeal to consumers' interests in such values as small-scale farming, humane conditions for livestock, beneficial working conditions for employees, or special protection for wildlife habitat or natural features of the region, such as lakes or rivers.

Regional Labels

The reputation of Dijon mustard, Roquefort cheese, and many other cheeses and wines of Europe has long been linked to their places of origin. Today in Europe there is a renewed interest in "marketing the region" through particular products, such as Comte Cheese in the Franche region of the French Jura mountain range, cheese in the Swiss Alps through the Pays d'En-Haurlin label, and specialty products under the Tastes of Anglia label in the East Anglia region of the United Kingdom (IATP 2000). These labels imply distinct flavors and quality or systems of processing unique to the region or community.

There was a time in the United States when almost every state had its own wine industry and produced apples, plums, potatoes, some of almost everything consumed in that region. As specialization and trade increased in the United States, people began to expect oranges and grapefruits to come from Florida, grapes and wine from California, potatoes from Idaho, and so forth. Gradually the market system became dominated by large food companies, producers, and processors, and consumers in the dairy state of Wisconsin soon knew more about Kraft cheese than they did the prizewinning cheddar made in their own county. Now with the North American Trade Agreement (NAFTA) and General Agreement on Tariffs and Trade (GATT), most consumers don't blink an eye when they buy grapes from Chile, apples from New Zealand, and strawberries from Mexico.

To renew an interest in regional food and keep more dollars for food circulating within an individual state, certain state departments of agriculture have started campaigns to encourage consumers to buy homegrown products. Labels such as "Minnesota Grown," and "Iowa Grown for You," can be found in regular supermarkets and appeal to the cultural loyalty of state residents. However, just five supermarket chains accounted for 42 percent of market share in the United States in 2000 (Hendrickson 2001). This kind of consolidation is increasing worldwide because it is more profitable to buy large quantities at special prices from a few suppliers, and loyalty to state products is irrelevant to the bottom line.

The closest link to place of origin are food items labeled or "branded" with the specific names of farms, such as Cedar Summit Farm and Earth-Be-Glad Farm in Minnesota. Customers go to the farm for their eggs or meat, though sometimes the farmer makes regular deliveries to nearby urban areas. This has worked well for livestock farmers who have access to a processing facility or "locker plant" in their community and can sell frozen meat. The major appeal to customers is that they can know where the food was produced and can talk to the farmer to find out how. They trust a local person, rather than a label (DeVore 1995).

This is true of Community Supported Agriculture (CSA) farms also. The first documented CSA farm was started in Massachusetts in 1985, and now there are an estimated one thousand across the country (Cone and Myhre 2000). Customers buy shares or memberships for a fixed price from a farm in their area and go to the farm or a pickup site near them for their weekly bags or boxes of vegetables during the growing season. Most CSAs are organic, though they may not have gone through the certification process. Being able to have vegetables straight from the field the day they were picked is more important to some consumers, especially when they know the farmer.

Cooperative Marketing

In recent years, small groups of producers have joined together to process and market their vegetables, meats, and dairy products collaboratively and sell them in their region, rather than be a part of the global market (King and DiGiacomo 2000). The most successful and well known is the Coulee Region Organic Produce Pool (CROPP), which started in the Upper Midwest with vegetables and milk and now processes and sells a variety of dairy products, eggs, and meat under the Organic Valley label (King and DiGiacomo 2000). The cooperative now also buys milk from farmers on the East and West Coasts to process and market regionally. The premium price Organic Valley pays for organic milk has made it economically possible for many dairy farmers to stay in the business.

Smaller-scale farmers who can't maintain a steady volume and flow of livestock to the big processors frequently get lower prices per unit than producers who can deliver semi-truck loads of livestock at a time, so they are experimenting with cooperative marketing. Cooperatives like the Prairie Farmers

Cooperative (hogs in western Minnesota) and the Rangeland Farmers Cooperative (chickens in southeast Minnesota) have been organized. Lack of sufficient capital to build or renovate processing plants and lack of experience or interest in business planning and marketing have been obstacles in the start-up of these efforts (King and DiGiacomo 2000). But even marketing cooperatively farmers can't compete with large-scale corporations. To get a higher price they need to "add value" that distinguishes their meat from meat in the conventional system. For example, products labeled "hormone free" or "antibiotic free" have an added value to the safe table shoppers, and they will pay more. Such products represent only one step toward environmental sustainability, but they do provide a fairer price for family farmers.

Cooperative marketing has worked well for a group of north-central Minnesota farmers. Every month, customers in the Twin Cities area can e-mail their order to the Whole Farm Cooperative for beef, pork, lamb, chicken, eggs, honey, and vegetables in season and then pick up their order at convenient delivery sites, such as offices and churches. Prices are higher than at supermarkets, but consumers are willing to pay for products direct from the farm. The mission of the Whole Farm Cooperative is "to create farms that nourish families spiritually and economically, sustain the environment, and provide customers not only with safe wholesome food but with a clear sense of who and where their food came from." Viewers can see photos of the farmers and read their farming philosophy on the web site. Meat bearing the Whole Farm Cooperative label is also available in some food cooperatives and small groceries in the Twin Cities. Customers trust that these farmers do use environmentally sensitive farming practices, but these claims are not verified by a third-party inspection process.

Eco-labels

"Green" labels are being introduced throughout the United States to appeal to environmentally concerned consumers. In some states, groups have established eco-labels that add value to the products by identifying environmental and social standards used by producers. Almost all of them have information and pictures on web sites. For example, the California Clean Marketing Group, formed in 1988, aims to strengthen California's small family farms, avoid toxic or environmentally degrading pesticides, and provide good conditions for workers (Nadeau 1998). One of the requirements to be certified for the label is that they must arrange their farms "in ways that encourage wildlife to take up permanent residence."

One eco-label that has captured a lot of attention in the Northwest is "Salmon Safe." The Pacific River Council's Salmon-Safe logo of two intertwined fish appears on select Northwest foods and beverages, including wines from some of Oregon's largest vineyards and rice from northern California. All certified farms are located in watersheds where salmon spawn and use practices that protect water quality. Salmon in the Northwest are charismatic species with a cultural identity that connects them to Native American history and the cuisine of the region. The public is quite amenable to paying higher prices for products grown by farmers who contribute to preservation of salmon habitat.

Another eco-label in the Northwest was created by the Food Alliance, a nonprofit organization formed in 1994 with a grant from the W. K. Kellogg Foundation to promote increased adoption of sustainable agriculture practices (Nadeau 1998). The consumer research completed prior to launching the label indicated that there was a significant, untapped market for Earth-sustainable products, but that most consumers did not include environmental or social considerations as part of their purchases (The Hartman Group 1997). This and other insights from the research guided the Oregon group in the introduction and establishment of its eco-label, Food Alliance Approved. After several successful years, it has become a model for other eco-labels.

The Midwest Food Alliance began as a joint project of the Cooperative Development Services, based in Wisconsin, and the Land Stewardship Project, based in Minnesota, to label foods produced sustainably in the Midwest. It started originally as an independent eco-label program called Food Choices, but later became a partner of the Food Alliance and adapted their already-tested environmental certification standards and processes. The Midwest Food Alliance eco-label was launched in the fall of 2000 with a promotion of certified apples in supermarkets (DeVore 2000). The Alliance is actively recruiting producers of fruits, vegetables, dairy products, and meats to join the program and become certified to use the seal, whether they market directly from the farm, at farmer's markets, or in retail stores.

The Food Alliance and Midwest Food Alliance are working to resolve two significant differences and strengthen the partnership. The Food Alliance is interested in marketing nationally under the Food Alliance Approved label, but the Midwest Food Alliance's goal is to market products primarily in the Midwest and build a regional food system. The Food Alliance has emphasized farm worker–farm owner relationships in developing its criteria for social sustainability, while the Midwest group has focused on family- and community-based farming. There is a possibility that current negotiations will produce a federation of regional food alliances and a blend of their different social criteria to evaluate social sustainability that will improve the eco-labels of both organizations.

Considering the many different kinds of labels and marketing experiments going on, most people who wish to provide a sustainable table in their homes now have opportunities to purchase appropriate foods, although they may be unaware of the available sources. However, consumers won't be able to learn about alternative markets and green-labeled products through the kind of costly television advertising used by the mainstream food system. Individuals and organizations who want farms to provide environmental benefits must help disseminate information about their products.

Building Market Demand

If we want to stop the trend toward mammoth industrial farms and create networks of farms that are natural habitats, the number of farms using sustainable systems must grow, and more people must buy for a sustainable table. We must greatly expand participation in alternative, regional food systems.

However, in writing about the potential of markets to fulfill the promise of organic agriculture, Allen and Kovac (2000) see limitations in trying to create change one consumer at a time. They write that collective problems can't be solved by the sum total of individual actions, which is the way the market works, but require collective action in the form of social movements. Environmental organizations and professional conservation groups should lead this movement to expand demand for sustainable food and sustainable agriculture. They are in a position to educate their constituencies about the connection between food and biological diversity and the connection between food choices for personal health and long-term health of the land. If all people with environmental sympathies—ranging from sportsmen protecting habitats for specific game birds to wilderness advocates protecting forests for their own sake, from people mainly concerned about environmental toxins that threaten human health to those concerned about the health of entire ecosystems—were to become avid consumers of foods with ecolabels, big changes could occur. Many farmers will be willing to alter their farming practices if it means higher prices for wildlife friendly products. Others who feel that they have been trapped into capital intensive industrial farming on too many acres will welcome an option to earn their living differently. And some will begin to renew a relationship between farming and the natural world that enhances the sustainability of both, while it improves the quality of life for the farmer and the community and society at large.

Ethics about recycling is part of the culture of environmental groups such as the Audubon Society and Sierra Club that has spread throughout society. Individuals are simply expected to recycle if they care about the environment. The message about buying food produced by sustainable methods could become just as pervasive as the message about recycling if consistently and repeatedly communicated through magazines, newsletters, web sites and meetings. Organizations could provide information about sources of local or regional food and help organize buying clubs.

Some environmental organizations, like the North Star Chapter of the Sierra Club, have been marketing shade-grown coffee, which requires fewer chemical inputs and provides natural habitat for birds. Bird-friendly or shade-grown coffee is also associated with social values, such as a fair price for the small coffee grower in Mexico, Guatemala, or Honduras and is sold under labels such as Fair Trade and Peace Coffee. The next thing these organizations could do is urge their members to demand cream and milk produced on grass-based, bird-friendly dairies in their region.

One positive influence in building the demand for local or green-label products is the Chef's Collaborative, a chef-managed national organization advocating sustainable cuisine. The Chefs describe themselves on their web site: "While celebrating the pleasures of food, Collaborative members recognize the impact of food on our lives, on the well-being of our communities and on the integrity of the global environment. And they celebrate the joys of local, seasonal, sustainable cooking."

The sustainable table must offer tasty foods that are convenient and easy to prepare, even if made from scratch, or it won't be a viable option. Family cooks must learn how to prepare a variety of fresh vegetables in season and how to cook more than a few cuts of meat. Many CSAs already put recipes in the weekly bags of vegetables, and meat producers are beginning to provide recipes also, but cookbooks and cooking classes for the sustainable table are needed. The fast-food table has conquered the family table not only because people haven't learned how to prepare healthful foods, but also because they have been told by its purveyors that they are too dumb or too important to take time to prepare a home-cooked meal.

Consumers who recognize "the impact of food on the integrity of the global environment," will find that buying and cooking for a sustainable table offers, in addition to more healthful food, the spiritual benefits of reconnecting with farmers, with the land, and with the ecosystem processes that food production depends upon.

A new organization that could have a unique influence on the movement for a sustainable food system is the Wild Farm Alliance, which was founded in 2000 by a group of wildlands proponents and ecological farming advocates. The wildlands advocates who work to restore large amounts of wildlands in the United States see agriculture as a primary cause of the global biodiversity crisis. They list habitat destruction and fragmentation, the displacement of native species and the introduction of exotic species, pollution of terrestrial and aquatic ecosystems, the persecution of predators, and the over-exploitation of nonrenewable resources for food production and distribution as a few of the ecological consequences of modern agricultural practices. But working with sustainable agriculture groups, the wildlands advocates hope to build a network of farmers, conservationists, and consumers all promoting a kind of farming that helps protect and restore wild nature. They envision a world in which "community-based, ecologically managed farms and ranches are seamlessly integrated into landscapes that accommodate a full range of native species and ecological processes." The consumer role in this vision is to understand the connection between food and biodiversity and to shop for a sustainable table.

Admittedly, the market today for sustainably produced foods is a niche market. Compared to sales by factory farms and giant food companies, it's a very small wedge of the economic pie. However, the potential for growth, and the impact growth could have on rural economies and landscapes, is enormous. Consumer demand for food produced sustainably and strong advocacy for a regional food system from all stripes of conservationists/environmentalists will take us a giant step closer to restoring a better relationship between agriculture and nature.

References

Allen, P., and M. Kovac. 2000. "The Capitalist Composition of Organic: The Potential of Markets in Fulfilling the Promise of Organic Agriculture." *Agriculture and Human Values* 17:221–232.

Chasteen, B. 1999. "Conscience, with a Price Tag." *Chronicle of Community* 3:15–26.

Cone, C. A., and A. Myhre. 2000. "Community Supported Agriculture: A Sustainable Alternative to Industrial Agriculture?" *Human Organization: Journal of the Society for Applied Anthropology* 59:187–197.

Council for Agriculture Science and Technology (CAST). 1999. *Benefits of Biodiversity*. Task Force Report No. 33. Council for Agriculture Science and Technology, Ames, Iowa.

DeLind, L. B. 2000. "Transforming Organic Agriculture into Industrial Organic Products: Reconsidering National Organic Standards." *Human Organization: Journal of the Society for Applied Anthropology* 59:198–208.

DeVore, B. 1995. "Locking in Local Markets." *Land Stewardship Letter* 13(6):12.

——-. 2000. "Midwest Food Alliance Label Launched in Minnesota." *Land Stewardship Letter* 18:5.

Grey, M. A. 2000. "The Industrial Food Stream and Its Alternatives in the United States: An Introduction." *Human Organization: Journal of the Society for Applied Anthropology* 59:143–150.

Hendrickson, M., W. D. Heffernan, P. H. Howard, and J. B. Heffernan. 2001. *Consolidation in Food Retailing and Dairy: Implications for Farmers and Consumers in a Global Food System.* National Farmers Union, Ames, Iowa.

Institute for Agriculture and Trade Policy (IATP). 2000. *Marketing Sustainable Agriculture: Case Studies and Analysis from Europe.* Institute for Agriculture and Trade Policy, St. Paul, Minn.

King, R., and G. DiGiacomo. 2000. *Collaborative Marketing: A Road Map and Resource Guide for Farmers.* Minnesota Institute for Sustainable Agriculture and University of Minnesota Extension Service, St. Paul.

Nadeau, E. G. 1998. *Marketing Sustainably Produced Foods: International Examples and Lessons for the United States.* Food Choices, Madison, Wis.

Pollan, M. 2001. "Supermarket Pastoral." *New York Times Sunday Magazine.* May 13.

The Hartman Group. 1997. *The Hartman Report. Food and the Environment: A Consumer's Perspective, Phase* 2. The Hartman Group, Bellevue, Wash.

Chapter 18

Agriculture as a Public Good

George M. Boody

The Koenen family farm is an island of grass in a sea of corn and soybeans. That's because in 1993 the Koenens—brothers Lyle and Paul, along with their father Ken—started converting row-crop fields to pasture. They did this when they adopted management intensive rotational grazing, a system that slashes the need for field crop-based feeds and all the expensive inputs that accompany them.

Such a livestock production system is a plus for rural communities reliant on retaining farmers and keeping them viable. Studies show that management intensive rotational grazing is an ideal way for beginning producers with few resources to get into farming and become profitable (Undersander et al. 1997). That means more businesses on Main Street, as well as more churches and schools. And because rotational grazing relies on replacing annual crops with perennial plant systems, it is also a boon to the environment. Fields covered with well-managed grass systems lose far less soil and other contaminants when compared with their row-cropped counterparts (Undersander et al. 1997, SFS in press). A farming system that reduces soil erosion and chemical runoff is a particularly valuable asset in this part of Minnesota: the watershed the Koenens farm in drains to the Minnesota River, which is one of the Mississippi River's leading sources of pollution (Mulla and Mallawatantri 1997).

The Koenens are typical of many U.S. farmers who are trying to make their operations more environmentally sustainable. They are the kind of farmers that society should be encouraging through federal farm programs. Instead, we have a federal commodity system that punishes such farmers for their creativity and diversification. After the Koenens decreased their acreage in corn, they were not eligible for as many government subsidies. They developed a grazing system despite farm programs that strongly encouraged them and their neighbors to maximize corn and soybean acreage and yields. Most of their neighbors won't make the changes the Koenens did, in part because they are unwilling to forego government payments in a time of volatile markets. Lyle estimates that in 1998 alone, their greener farming system cost the operation $3,000 in lost government payments. Such a penalty hurts financially at a time of extremely thin profit margins. But just as important, it sends a clearly negative message: diverse farming based on perennial plant systems is not what society wants.

Farmers and ranchers own and manage 50 percent of the land area in the United States, 907 million acres of cropland, pastureland, and rangeland (NRCS 1996). Farm policies designed in Washington, D.C., affect the land use decisions they make. In turn, these decisions affect the health and diversity of the natural environment on rural landscapes, the quality of life for people living in rural communities, both now and in the future, and the kind of food urban people eat.

What has the public gained from federal farm policy? Not much. Citizens have paid staggering amounts of tax dollars in farm income to some of the country's largest farmers to produce surpluses of only a few commodities, which are then available at low prices to traders and processors. In a three-year study of the U.S. farm commodity subsidy system, using computer modeling, the World Resources Institute concluded that under such a system, "tax dollars are being wasted twice: once on crop subsidies and again to clean up environmental damage caused by subsidized farming practices" (Faeth 1995).

Clearly the basic framework of agricultural policy no longer makes sense for small- to mid-sized farmers, for the land, or for the public. In this chapter, I will review the income and commodity subsidy programs that largely drive conventional agriculture and some of the conservation programs intended to mitigate environmental problems caused by commodity production. Then I will propose a new framework for agricultural policy based the idea that agriculture can be a public good and farms can and should produce multiple benefits for society.

This book is about ending the war that industrial agriculture wages against nature and restoring a harmonious relationship between farming and the natural world. We must turn away from the failed farm programs that continue to subsidize this war and build a new base for farm support payments that will make agroecololgical restoration possible.

Farm Policy 101

For seventy-some years, beginning in the 1930s, government commodity policies have assured production of a relatively small number of crops. Just seven program crops—corn, wheat, sorghum, barley, oats, rice, and cotton—received direct subsidies until the recent addition of soybeans and minor oilseed crops (Hoffman 1999).

A farmer's eligibility for subsidies for any of the program crops before 1996 was usually based on the average number of acres planted in the past and average yields of the crop (NRC 1989). This was known as "base acres." No income tests were required; the more farms produced through intensive use of chemicals and cultivation, the more lucrative the farm program became (Levins 2000). It is estimated that taxpayers spent $291 billion dollars in subsidy programs between 1978 and 1994 (Janick et al. 1996). That averages out to about $17 billion a year.

This system was supposed to change in 1996 when Congress passed the sweeping Federal Agricultural Improvement and Reform Act—known most commonly as the "Freedom to Farm Act." Ostensibly, it was to have changed federal farm commodity policy to separate the amount of farm payments from the kinds and amount of crops grown. The featured component of the program was a subsidy for farmers linked to the number of acres they had enrolled in past farm programs but not linked to particular crops, so farmers could plant whatever they wanted. The payments were to be gradually decreased to zero over seven years (Hoffman 1999). The federal government provided $22.9 billion of income for farmers enrolled in this production flexibility program between 1996 and 1998 (EWG 2000).

Initially, environmental groups and sustainable agriculture advocates were cautiously optimistic about the Freedom to Farm Act. In theory, farmers could choose what to grow based on markets, rather than what the government would pay for, and could diversify their operations. However, in reality, this revamped farm policy may have made things even worse.

During the late 1990s, farmers experienced extraordinarily low commodity prices. To help farmers out, Congress gave them billions of dollars in additional "emergency" and "bailout" money. For example, farmers were eligible to receive a loan deficiency payment, which was directly tied to how many acres of program crops they grew. Seven crops—barley, oats, canola, sunflower, upland cotton, and rice—received fifteen percent of $6.78 billion in loan subsidies. Just two crops, soybeans and corn, accounted for $4.74 billion of those dollars (Williams-Derry and Cook 2000). No wonder more corn and soybeans are being grown than ever

(Price 2001). Granted, the Freedom to Farm Act has resulted in some "crop shifts"—more soybeans are being grown in western states, for example—but it has fallen far short of creating anything approaching diversity.

Much of this money went to support a few farmers. The top 10 percent of farmers who received loan subsidies, or about 88,000 farms, received 56 percent or $3.77 billion in 1999. Congress also doubled program payments and boosted crop insurance payments. It lifted caps on loan subsidies for the largest grain producers, which benefited primarily 3,400 large individual or farm businesses (Williams-Derry and Cook 2000). Ironically, the average taxpayer thought this money was going to help family-sized farming operations.

The Freedom to Farm Act has not lived up to its promise to either cut government payments or promote flexibility. In fact, the policy has cost a bundle and has driven agriculture strongly in the direction of less crop diversity. According to a *New York Times* article on December 24, 2000, the government spent $28 billion on the farm programs in 2000 alone. That makes the $17 billion per year being spent before the Freedom to Farm Act look like a bargain.

It's not just large, undiversified grain farmers who win under this policy. Inadvertently, the Freedom to Farm Act has resulted in a subsidy for large-scale livestock producers. Because it encourages overproduction, this policy keeps prices of feed grains well below what it costs a typical farmer to produce them. This puts smaller livestock producers, who raise their own feed, at a competitive disadvantage (Specht 2001). Thus, again, diversity is penalized.

It could be argued that when the Freedom to Farm Act expires in 2002, the free market will take over agriculture, and public coffers will no longer be opened wide for the sake of a few large farmers. It's doubtful the government's involvement in agriculture will end. But even if that were to happen, society will be paying the environmental costs created by this subsidy system long into the future.

Conservation and the Farm Bill

To be fair, federal farm policy has made an attempt to address the environmental costs of agriculture since 1935. More recently, programs that do everything from providing cost-share funds for farmers who voluntarily establish environmentally friendly structures or practices to funding outright land retirement have proven popular within the farming and environmental communities. In terms of overall spending, conservation programs are only funded at about one-half of their 1937 level (NRCS 1996). In 2000, they were only about 8 percent of the total federal outlay for farms (Green and McElroy 2001). But underfunding is only part of the problem with these programs. They operate under the philosophy that agroecological problems can be dealt with only by leaving the land completely alone or by prescribing a defined set of practices outlined in technical manuals.

For example, the U.S. Department of Agriculture's Environmental Quality Incentives Program (EQIP) relies on the use of best management practices (BMPs) to attain environmental goals. Best management practices are practices or handfuls of practices designed for the most part to solve an isolated environmental problem, such as nutrient runoff, within the context of a dominant row crop or animal feedlot system. Instead of considering whether there are viable alternatives to raising corn or soybeans on a highly erosive piece of land, best management practices are typically used to provide a way to make conventional farming systems more environmentally palatable. Under the Environmental Quality Incentives Program, qualified farmers are paid to partially offset the cost of adopting specified approved practices. One example of a government-approved BMP is "no till" or "minimum till," which consists of planting seeds directly into the previous year's crop stubble using specially designed tillage equipment. This eliminates plowing and reduces erosion by leaving the land covered with dead plant material. But less tillage often means more herbicide use to control weeds (Gebhardt et al. 1985). So, one environmental problem—in this case soil erosion—is dealt with, but another one—more toxins in the environment—is created or exacerbated. Of course, best management practices funded by EQIP do not always result in such direct environmental tradeoffs. For example, perhaps soil erosion is so bad in a particular area that putting more herbicides in the environment is worth the risk. And a small percentage of total EQIP funds do go to farmers making the change to management intensive rotational grazing and other sustainable systems that take into account a big picture view of the farm. But too often best management practices are prescribed and applied in isolation from landscape-level ecological issues.

Farm policy has used variations of the "stick" approach to achieve environmental goals as well. For example, Conservation Compliance was implemented in 1985. This program required farmers to adopt certain conservation practices—leaving more residue on their soil after harvest, for example—in order to stay eligible for commodity payments. The more highly erosive land was, the tougher the requirements. To a certain extent, it worked. Erosion on cropland was cut by nearly 38 percent between 1982 and 1997. However, beginning in 1995, erosion reductions leveled off. Whatever the reason, 29 percent of our crop fields are "excessively eroding," say government soil experts (NRCS 2001). And farmers tell me that government officials are making fewer site visits to ensure compliance.

Over 70 percent of overall conservation program dollars spent since 1988 have been for land taken out of production (Claassen et al. 2001). Most of that money went to the Conservation Reserve Program, also known as CRP, which pays farmers to voluntarily remove vulnerable lands from row crop production for a ten-year period and requires permanent cover to be established. As many as 36 million acres have been enrolled. As originally envisioned, it was a response to the major overproduction of the early 1980s and significant erosion levels from bringing marginal lands into production during the boom years of the 1970s. It was designed to function alongside commodity programs that continued to support increasing yields on the remaining acres of commodity crops.

The Conservation Reserve Program has reduced soil erosion and produced wildlife benefits. The U.S. Department of Agriculture estimates that total erosion on CRP land fell by about 406 million tons per year between 1982 and 1997. Wildlife biologists say game birds such as pheasants in particular have benefited from the increased habitat produced by CRP set-asides. Overall, the Conservation Reserve Program has pumped an extra $704 million dollars per year into rural economies as a result of the increased opportunities for wildlife viewing and pheasant hunting it has provided (Claassen et al. 2001).

These are some impressive benefits. However, they pale a bit if we deduct erosion and habitat loss that resulted from converting hay or pasture lands back into row crop production as a result of the Conservation Reserve Program (Claassen et al. 2001). This happened because land retirement helped reduce the commodity program surpluses early on and commodity prices rose. Higher prices induced farmers to convert their hay and pasture lands into row crops. CRP acres were protected, but other acres suddenly became vulnerable because the program was so effective at reducing surpluses.

Land set-asides have their place, particularly in ecologically sensitive areas that just shouldn't be farmed. But widespread land retirement also ignores the kind of true ecological benefits that can be attained by sustainable farmers who are making a living on the land. Once those practices are established and become viable within a farmer's management system, congressional policies are less likely to uproot them.

Good Agriculture/Multiple Goods

We shouldn't be surprised that initiatives such as the Conservation Reserve Program, the Environmental Quality Incentive Program, and Conservation Compliance are limited in their ability to create an ecologically sound farming system. There is only so much tweaking that can be done to make an inherently flawed policy more sustainable. U.S. farm policy has its foundation in a narrowly focused philosophy that sees commodities as the overriding benefit to come from farming. Such thinking produces policy that, no matter how much it is modified, will still produce more cheap grain at the expense of the agroecosystems. Best management practices, and land retirement to a certain extent, provide a handy crutch for row crop production to continue. Farmers and others interested in creating agroecosystems that are more resilient say policy must be developed that recognizes the multiple benefits or public "goods" farms can provide beyond bulging grain bins. Public goods are those benefits society deems it needs but does not directly pay for by the exchange of goods and money through the marketplace.

There are plenty of goods that don't appear on any label but that people value all the same: aesthetic landscapes, songbird and waterfowl habitat, carbon capture, and community jobs. A public good can also take the form of removing or avoiding the public "bads" currently created by industrialized agriculture: contaminated drinking water, polluted streams, reduced wildlife populations, and increased lung disease problems produced by working conditions in livestock confinement. But how do we create incentives to provide these goods?

Ultimately, American consumers may have to support public goods by paying farmers for them. Currently, notes farmer and sustainable agriculture pioneer Fred Kirschenmann (2001), agriculture is seen more as a public problem than a public good. This must change. We need a new farm policy that clearly promotes nonmarket benefits as well as market benefits. This policy should provide a clear and strong signal to farmers that society wants them to manage their land in a way that optimizes the balance between production of crops and livestock, and public benefits.

The idea that agriculture can provide public goods is not new. Agricultural economist and former presidential adviser Willard Cochrane likens the food production and distribution system to the education and health care systems. He considers the products of those systems basic human rights and believes that public governance is essential to regulate such systems (Levins 2000). Economists Douglas Gollin and Melinda Smale (1998) argue that agriculture itself would benefit from certain "thresholds of on-farm diversity" that pure market forces are not equipped to provide. It is not in the interest of Cargill, ConAgra, or Archer Daniels Midland to pay farmers for increased diversity. However, consumer preference studies show that the average American does think such a "good" is worth paying for.

One such study was part of an initiative the Land Stewardship Project helped launch in 1999, the Multiple Benefits of Agriculture Project. A statewide survey of citizens showed that Minnesotans are willing to pay increased taxes or marketplace costs of $201 per household over a ten-year period for significant, environmentally friendly practices in agriculture. That translates into $362 million per year in Minnesota alone (Krinke and Boody 2001). If that figure were true for all states, payments for public goods would equate to the $17 billion level of payments for commodity programs.

The Multiple Benefits of Agriculture Project looked at two watersheds, the Chippewa River watershed in the gently rolling prairie farms of western Minnesota and the Wells Creek watershed in hilly southeastern Minnesota. Project participants developed four land management scenarios: (1) extension of current agricultural land use, (2) the adoption of best management practices in existing cropping systems, (3) the adoption of more diverse, integrated farming systems, and (4) adding significant perennial cover through prairie, pastures for rotational grazing, and 300-foot-wide buffer strips along streams. Through modeling, they learned that diversifying agricultural production systems in these two watersheds could dramatically increase the amount and variety of public goods. In the Wells Creek watershed with steeper slopes, the amount of sediment, phosphorous, and nitrogen entering the stream could be reduced by as much as 80 percent by increasing crop diversity, using best management practices on crop land, adding more grassland buffers, and pasturing more animals on the land instead of in feedlots. (BMPS by themselves would reduce nitrogen by only 40 percent.) The model showed that the resulting sediment reduction would decrease predicted fish kill events by one-half. Maps produced in a geographic information system clearly show how more buffer areas along streams would connect and form corridors of perennial vegetation that could benefit wildlife. In the flatter Chippewa watershed, modeling showed that fully utilizing best management practices in existing cropping systems could reduce in-stream nitrogen levels 18 percent but that adding small grains and hay in significant amounts to the land-use mix could reduce in-stream nitrogen levels 60 percent. Scientists generally agree that a reduction of 40 percent in the Mississippi watershed is necessary to significantly shrink the hypoxia zone in the Gulf of Mexico.

The recognition that agriculture produces more than crops and livestock is understood internationally as the "multi-functional nature of agriculture" (see chapter 10). An example of government policy to support multifunctional agriculture is the national Agricultural Orientation Law (Loi d'Orientation Agricole) in France, which is based on local community involvement and regional goals that are harmonized with national and European law. A farmer can enter into a "Rural Farming Contract" (Contrats Territoriaux d'Exploitation) with the government at the departmental level based on commitments he or she is willing to make. For creating added value such as "quality improvement of products, farming diversification, incentive to create and maintain jobs," and improving land management that affects water, grasslands, biodiversity, and the like, the farmer receives government payments. The Agricultural Orientation Law requires the French government to develop structural policies that maintain farms "of a reasonable size" and are independently operated, and it also includes support for assisting beginning farmers and other measures to keep a large number of persons active in farming (Vorley 2001).

More Than Just Tweaking

The United States needs to develop policy that treats agriculture as a national public good. Agricultural production is inherently local, but its inputs and outputs cross many ecosystem boundaries. For example, tractor fuel is trucked into the Chippewa watershed, and Chippewa watershed silt flows over the Chippewa County line, and over the Minnesota border, for that matter. The Koenens farm near the Chippewa River, which is a tributary to the Minnesota and Mississippi Rivers. That means any contaminants the Koenens produce have the potential to make it to the Gulf of Mexico. We need to strike a balance between allowing local people to deal creatively with local problems while producing results that have positive results downstream or across borders.

To do so, we need minimum environmental standards at the national level. This is a new concept for agriculture (Runge 1997). A detailed analysis of U.S. environmental policy as applied to agriculture found that with the exception of the Endangered Species Act, Congress has mostly exempted farms from regulation. For example, policies related to the Clean Water Act outline potential minimum standards such as total maximum daily load standards

(TMDLs) for impaired watersheds across the country. But states can decide whether they will apply TMDLs to farms within their borders (Ruhl 2001).

A main goal of U.S. agricultural policy has been to provide abundant, low-cost food to its citizens. Now the country needs an equally important goal: to provide food safety and quality associated with production systems. For example, current government quality indices are biased toward beef raised in feedlots, which accounts for 78 percent of all beef production in the United States (Conner et al. 1999). The system of feedlot-raised beef causes the biggest environmental problems in the cattle industry. Quality indices should be changed to recognize that grass-fed meat, grown on perennial pastures and ranges with few environmental problems, is lower in cholesterol and fat (Conner et al. 1999) and thus is a more desirable product. Studies also indicate that grass-fed beef is less likely to pose an *Escherichia coli* (*E. coli*) threat to people who eat it (Diez-Gonzalez et al. 1998). The U.S. Department of Agriculture should recognize these benefits and make them known to retailers and consumers.

As we've seen in previous chapters, sustainable agriculture is a management and information-intensive enterprise. If we expect deep changes to occur on our farms, a national policy must be developed that significantly increases technical assistance, research, and education in sustainable agriculture for farmers (NCSF 1998).

A good model for providing farmers with technical assistance is the Department of Agriculture's Sustainable Agriculture Research and Education Program (SARE). This unique program has directed research dollars—$13 million in 2000 (Vorley 2001)—into sustainable agriculture, and it has done so through the participation of diverse stakeholders. The program is administered in four regions by regional councils. The North Central SARE council, for example, includes farmers and representatives from academia, extension, nonprofit organizations, and agribusiness. The program is popular among farmers because it emphasizes on-farm research. Scientists and environmentalists like it because such research often produces solid, real-world results.

The French Agricultural Orientation Law and SARE provide clues as to how we can reach national goals primarily through local initiatives. Federal policy should allow for local input and site specificity so that private landowners can adopt farming systems appropriate to their geographical area. It should include programs to reimburse farmers for changing farming systems, retiring land where needed, and provide income supplements to farmers who produce public goods.

Who Should Get Subsidized?

The question of which farmers should get subsidies is a sticky one. Farmers who already use stewardship-based systems should have access to the income provisions also, not just those who are changing systems to meet the new standards. However, some policy analysts contend that ecologically sound farming systems that can be adopted without imposing a major economic burden on farmers should be maintained in the course of doing business, without government subsidies. Some farming systems—organic production for example—can prove profitable in the free market once it is established. On the other hand, crop farming without artificial drainage is a plus for wildlife and water quality but can be a financial burden for farmers. Some policy analysts note that farm income support programs are viewed as entitlements by recipients, and the expected income streams are regularly capitalized into land values. It will be important to find effective ways to decouple publicly funded incentives from land values (Dobbs and Pretty 2001). How should society determine which farmers get subsidized? Let's look at two recent examples of policy mechanisms that hold promise for making such a decision in a fair and productive manner: the Resource Conservation Agreement and the Farm Results Index.

Resource Conservation Agreement

The Florida Stewardship Foundation developed the Resource Conservation Agreement to fill gaps left by existing conservation programs. The voluntary agreement between a private landowner and a government or nonprofit organization would last for at least twenty years and cover the implementation of a plan to protect "landscapes that harbor endangered species, wetlands, or other natural ecosystems." In return, the farmer could receive tax breaks or other conservation funding, as well as assurances that the economic use of the land could continue. An important side benefit is the opportunity for participating farmers to develop trust with local land management agencies or groups. A pilot project has been proposed in Florida to funnel $46 million in payments to landowners over 170,000 acres to protect panther habitat and other natural resource values. This type of agreement is also being explored in Oregon and California (FSF 2001). It could be expanded as an income payment program.

Farm Results Index

The Land Stewardship Project, in consultation with farmers throughout the Midwest, has developed the Farm Results Index. To determine this index, points would be awarded to farms according to environmental and social results they achieved as defined by national and regional goals. This system is not without precedent: the U.S. Department of Agriculture uses an Environmental Benefits Index to review eligibility for the Conservation Reserve Program. However, the Farm Results Index would include biodiversity, wetland protection, ground and surface water protection, and social factors. It would also be recalculated every year.

Under such a program, taxpayers save twice. First, they won't pay farmers to produce crops in ways that create problems. Second, the money only goes to farmers who show demonstrable environmental results. This second point is important. Currently, government programs rely on measuring the change in practices, such as putting in a terrace, as a proxy for environmental improvement, or require no check at all. The new approach will require the emphasis to be shifted to estimating, observing, and, in some cases, measuring results.

A Farm Results Index will need to be administratively feasible (which also means feasible for farmers), cost effective, and replicable. It may be useful to utilize modeling and valuation of selected nonmarket benefits as a way to assess progress (Bockstael et al. 1995, Pretty et al. 2001). An important step is to identify indicators. For example, the Monitoring Team (see chapter 6) developed biological, social, and financial indicators that can be used by farmers and other professionals to observe trends on the farm. Richard Levins, a University of Minnesota economist who worked extensively with the Monitoring Team, developed a set of four financial indicators of sustainability: reliance on government programs, use of equipment, chemicals and nonrenewable energy, creation of jobs, and balance between feed use and feed production. Levins has set up a simple "calculator" to figure the sustainability of an operation based on accounting numbers that farmers must have to do their income taxes.

The calculator can produce some surprising results. For example, based on actual farm records, Levins ran four Minnesota operations through the calculator: a conventional corn-soybean farm, a conventional grain operation that finishes hogs, a conventional dairy, and a grass-based dairy. What he found was that although a 960-acre farm growing only corn and soybeans may produce a reasonably good income for its owners, it is a "complete failure" in terms of the financial indicators of sustainability. He noted that "few other local people share in the benefits. There is no chance for animals to work on the farm. And the farm ties up enough land to support at least three farms using different methods." On the other hand, a grass-based dairy that has 250 acres of carefully managed pasture and no corn was able to produce a good income for the owners, as well as produce jobs in the community, Levins reported. In addition, the grazing system required virtually no agricultural chemicals, which is seen as a major benefit to the environment (Levins 1996).

In early 2001, a policy proposal that could go a long way toward reaching these goals was introduced in Congress. The Conservation Security Act would include graduated payments depending on the extent of benefits available from the farm. A landowner would get the smallest number of dollars for implementing best management practices and a higher level for including more diverse farming systems that address multiple benefits through whole farm resource plans. This conservation program would provide funds to anyone who qualified, without regard to their cropping history (Harkin 2001).

Conclusion

There is no one silver-bullet policy that will suddenly make our farm land an ecological paradise. I mention proposals such as the Conservation Security Act only to show that some of the principles of "agroecological restoration" and other "multiple benefits" or public "goods" are starting to take hold in the minds of agricultural decision makers. Farmers, environmentalists, and others are starting to get the message across that farm policy must consist of more than "a dollar more for corn."

It is time for even more public dialogue in the United States on what is needed to harvest public goods from farming. Agricultural policy must not be left to negotiations between powerful politicians in Congress protecting the status quo and the lobbying arms of farm and commodity groups that primarily represent the largest farmers who reap the bulk of public income support. Instead, the public at large must be involved in helping determine national and local goals for agriculture, thereby contributing to a new vision and social contract for agriculture.

References

Bockstael, N., R. Costanza, I. Strand, W. Boynton, K. Bell, and L. Wainger. 1995. "Ecological Economic Modeling and Valuation of Ecosystems." *Ecological Economics*. 14:143–159.

Claassen, R., L. Hansen, M. Peters, V. Breneman, M. Weinberg, A. Cattaneo, P. Feather, D. Gadsby, D. Hellerstein, J. Hopkins, P. Johnston, M. Morehart, and M. Smith. 2001. *Agri-Environmental Policy at the Crossroads: Guideposts on a Changing Landscape*. Agricultural Economic Report No. 794. U.S. Department of Agriculture, Economic Research Service, Washington, D.C. Available at http://www.ers.usda.gov/publications/aer794.

Conner, J. R., R. A. Dietrich, G. W. Williams. 1999. *The U.S. Cattle and Beef Industry and the Environment.* A Report to the World Wildlife Fund. May.

Diez-Gonzalez, F., T. R. Callaway, M. G. Kizoulis, and J. B. Russell. 1998. "Grain Feeding and the Dissemination of Acid-Resistant *Escherichia coli* from Cattle." *Science* 281:1666–1668.

Dobbs, T. L., and J. N. Pretty. 2001. "The United Kingdom's Experience with Agri-environmental Stewardship Schemes: Lessons and Issues for the United States and Europe." Staff Paper 2001-1. South Dakota State University, Brookings.

Environmental Working Group (EWG). 2000. "Green Acres: How Taxpayers Are Subsidizing the Demise of the Family Farm. April 2000." Electronically published by the Environmental Working Group. Available at http://www.ewg.org/pub/home/reports/greenacres/exec.html.

Faeth, P. 1995. *Growing Green: Enhancing the Economic and Environmental Performance of U.S. Agriculture.* World Resources Institute, Washington, D.C.

Florida Stewardship Council (FSF). 2001. The Farm Stewardship Program. A Series of Reports from the Florida Stewardship Foundation. Boca Raton, Fla. Available at http://www.privatelands.org/RCA_open.htm.

Gebhardt, M. R., T. C. Daniel, C. E. Schweizer, and R. R. Allmaras. 1985. "Conservation Tillage." *Science* 230:625–630.

Gollin, D., and M. Smale. 1998. "Valuing Genetic Diversity: Crop Plants and Agroecosystems." Pp. 243–244 in *Biodiversity in Agroecosystems*, edited by W. W. Collins and C. O. Qualset. CRC Press, Boca Raton, Fla.

Green, R., and R. McElroy. 2001. "Farm Income and Costs: Farm Income Forecast." U.S. Department of Agriculture, Economic Research Service. Available at http://www.ers.usda.gov/Briefing/FarmIncome/fore.htm.

Harkin, T. 2001. Statements on Introduced Bills and Joint Resolutions. S.J. Res. 932. 107th Cong., 1st sess. Cong. Rec., May 22.

Hoffman, L. 1999. "Title I: Agricultural Market Transition Act." Pp. 3–29 in *Provisions of the Federal Agriculture Improvement and Reform Act of 1996*, edited by F. J. Nelson and L. P. Schertz. *Agricultural Information Bulletin* 729. U.S. Department of Agriculture, Economic Research Service. Available at http://www.ers.usda.gov/publications/aib729.

Janick, J., M. G. Blase, D. G Jolliff, and R. L Myers. 1996. "Diversifying U.S. Crop Production." Pp. 98–109 in *Progress in New Crops*, edited by J. Janick. ASHS Press, Alexandria, Va.

Kirschenmann, F. 2001. *Questions We Aren't Asking in Agriculture: Beginning the Journey Toward a New Vision.* Presentation at Foodsheds, Multi-Functional Agriculture and Fair Trade Conference, April 22, Pittsburgh, Penn.

Krinke, M. and G. Boody. 2001. "The Multiple Benefits of Agriculture: An Economic, Environmental and Social Analysis." A report from the Multiple Benefits of Agriculture Project. Land Stewardship Project. White Bear Lake, Minn. Available at www.landstewardshipproject.org.

Levins, R. A. 2000. *Willard Cochrane and the American Family Farm.* University of Nebraska Press, Lincoln.

Levins, R. A. 1996. *Monitoring Sustainable Agriculture with Conventional Financial Data.* Land Stewardship Project, White Bear Lake, Minn.

Mulla, D. J., and A. P. Mallawatantri. 1997. *Minnesota River Basin Water Quality Overview*. FO-7079-E. University of Minnesota Extension Service, St. Paul.

National Commission on Small Farms (NCSF). 1998. *A Time to Choose: A Report of the National Commission on Small Farms*. Miscellaneous Publication 1545. U.S. Department of Agriculture, Washington, D.C.

National Research Council (NRC). 1989. *Alternative Agriculture: Committee on the Role of Alternative Farming Methods in Modern Production Agriculture*. Board on Agriculture, National Research Council. National Academy Press, Washington, D.C.

National Resource Conservation Service (NRCS). 1996. *America's Private Land: A Geography of Hope*. U.S. Department of Agriculture, Natural Resources Conservation Service, Washington, D.C.

——. 2001. *Natural Resources Inventory: Highlights*. January. U.S. Department of Agriculture, Washington, D.C.

Pretty, J., C. Brett, D. Gee, R. Hine, C. Mason, J. Morison, M. Rayment, G. Van Der Bijl, and T. Dobbs. 2001. "Policy Challenges and Priorities for Internalizing the Externalities of Modern Agriculture." *Journal of Environmental Planning and Management*. 44:263–283.

Price, G. K. 2001. Field Crops: "Large Field Crop Supplies Expected Again in 2001/02." *Agricultural Outlook*.AGO-272. June-July. U.S. Department of Agriculture, Economic Research Service, Washington, D.C. Available at http://www.ers.usda.gov/publications/agoutlook/jun2000/contents.htm#one.

Ruhl., J. B. 2001. "The Environmental Law of Farms: Thirty Years of Making a Mole Hill Out of a Mountain." *The Environmental Law Reporter: News and Analysis* 31:10203–10223.

Runge, C. F. 1997. "Environmental Protection from Farm to Market." Pp. 200–216 in *Thinking Ecologically: The Next Generation of Environmental Policy*, edited by M. R. Chertow and D. Esty. Yale University Press, New Haven.

Specht, D. 2001. "Farm Policy's Three Fatal Flaws Threaten U.S. Competitiveness." *Land Stewardship Letter* 19(2):2–3.

The Sustainable Farming Systems Project (SFS). In press. University of Minnesota, Minnesota Institute for Sustainable Agriculture, St. Paul.

Undersander, D., B. Alber, P. Porter, A. Crossley, and N. Martin.1997. *Pastures for Profit: A Guide to Rotational Grazing*. Publication A3529. University of Wisconsin Cooperative Extension Service. Madison.

Vorley, W., 2001. "Farming That Works: Reforms for Sustainable Agriculture and Rural Development in the EU and US." Background paper to Sharing Responsibilty for Promoting Sustainable Agriculture and Rural Development: The Role of EU and US Stakeholders Transatlantic Workshop. January 24–26, Lisbon, Portugal. Revised version March 2001.

Williams-Derry, C., and K. Cook. 2000. "Bumper Crop: Congress' Latest Attempt to Boost Subsidies for the Largest Farms." Electronically published by the Environmental Working Group. Available at http://www.ewg.org/pub/home/reports/bumpercrop/execsumm.html.

About the Contributors

Collin A. Bode is a graduate student at the University of California, Santa Cruz.

George M. Boody is the executive director of the Land Stewardship Project and director of its Multiple Benefits of Agriculture Program.

Nina Leopold Bradley is on the board of directors for the Aldo Leopold Foundation and the editorial board of directors for *Orion Quarterly*.

Brian A. DeVore is the communications coordinator for the Land Stewardship Project and editor of its principal publication, *The Land Stewardship Letter*.

Arthur S. (Tex) Hawkins works for the U.S. Fish and Wildlife Service as the watershed biologist for the Upper Mississippi River refuges in the blufflands of Minnesota, Wisconsin, Iowa, and Illinois.

Wellington (Buddy) Huffaker is the executive director of the Aldo Leopold Foundation.

Dana L. Jackson is the associate director of the Land Stewardship Project.

Laura L. Jackson is an associate professor of biology at the University of Northern Iowa where she teaches courses in conservation biology, applied ecology, and environmental studies.

Rhonda R. Janke is an associate professor of horticulture and extension specialist at Kansas State University.

Richard G. Jefferson works on problems related to the conservation, ecology, and management of semi-natural grasslands in the English lowlands for English Nature, the British government's advisor on nature conservation in England.

Nicholas R. Jordan is a crop/weed ecologist in the Department of Agronomy and Plant Genetics at the University of Minnesota.

Cheryl Miller works for the National Audubon Society where she directs policy and educational efforts on Minnesota's wetlands and watersheds.

Heather J. Robertson is a national specialist on problems related to the conservation, ecology, and management of semi-natural grasslands in the English lowlands for English Nature, the British government's advisor on nature conservation in England.

Carol Shennan is the director of the Center for Agroecology and Sustainable Food Systems and a professor in the Environmental Studies Department at the University of California, Santa Cruz.

Judith D. Soule is the director of the Michigan Natural Features Inventory, a program of Michigan State University Extension Service.

Beth E. Waterhouse is a writer and consultant to nonprofit organizations and foundations in Minnesota. She teaches environmental ethics at the University of Minnesota, Department of Fisheries and Wildlife.

Index

Island Press
Board of Directors